Essential
Nuclear Medicine Physics

To my two nuclear families: Ronald, Arianna, and Danny, and Edward, Rhoda, Seth, Ethan, and David, for their love and support.

RAP

To Rhoda M. Powsner, M.D., J.D. for her love, support, and her continuing help.

ERP

SECOND EDITION

Essential Nuclear Medicine Physics

Rachel A. Powsner

M.D.
Associate Professor of Radiology
Boston University School of Medicine
Director, Division of Nuclear Medicine
Department of Radiology
Boston Veterans Administration Healthcare System
Boston, Massachusetts

Edward R. Powsner

M.D.
Former Chief, Nuclear Medicine Service, Veterans Administration Hospital
Allen Park, Michigan
Former Professor and Associate Chairman, Department of Pathology
Michigan State University
East Lansing, Michigan
Former Chair, Joint Review Committee for Educational Nuclear Medicine Technology
Former Member, American Board of Nuclear Medicine

Blackwell
Publishing

© 2006 Rachel A. Powsner and Edward R. Powsner
Published by Blackwell Publishing Ltd

Blackwell Publishing, Inc., 350 Main Street, Malden, Massachusetts 02148-5020, USA
Blackwell Publishing Ltd, 9600 Garsington Road, Oxford OX4 2DQ, UK
Blackwell Publishing Asia Pty Ltd, 550 Swanston Street, Carlton, Victoria 3053, Australia

The right of the Author to be identified as the Author of this Work has been asserted in
accordance with the Copyright, Designs and Patents Act 1988.

First edition published 1998
Second edition published 2006
1   2006

Library of Congress Cataloging-in-Publication Data

Powsner, Rachel A.
Essential nuclear medicine physics/Rachel A. Powsner, Edward R. Powsner. – 2nd ed.
p.; cm.
Rev. ed. of: Essentials of nuclear medicine physics. 1998.
Includes index.
ISBN-13: 978-1-4051-0484-5 (alk. paper)
ISBN-10: 1-4051-0484-8 (alk. paper)
1. Nuclear medicine. 2. Medical physics. I. Powsner, Edward R., 1926-. II. Powsner,
Rachel A., Essentials of nuclear medicine physics. III. Title.
[DNLM: 1. Nuclear Medicine. 2. Accidents, Radiation – prevention & control. 3. Nuclear
Physics. 4. Radiation Effects. 5. Radiation.
WN 440 P889e 2006]
R895.P69 2006
616.07′575–dc22
2005035905

A catalogue record for this title is available from the British Library

Set in 9/13pt Palatino by Newgen Imaging Systems (P) Ltd., Chennai, India
Printed and bound in Replika Press Pvt Ltd, Haryana, India

Commissioning Editor: Martin Sugden
Development Editor: Lauren Brindley
Production Controller: Kate Charman
Editorial Assistant: Eleanor Bonnet

For further information on Blackwell Publishing, visit our website:
www.blackwellpublishing.com

The publisher's policy is to use permanent paper from mills that operate a sustainable
forestry policy, and which has been manufactured from pulp processed using acid-free
and elementary chlorine-free practices. Furthermore, the publisher ensures that the text
paper and cover board used have met acceptable environmental accreditation standards.

Contents

Preface

After years of postgraduate training, many physicians have forgotten some (or most) of their undergraduate and high school physics and may find submersion into nuclear physics somewhat daunting. This book begins with a very basic introduction to nuclear physics and the interactions of radiation and matter. It then proceeds with discussions of nuclear medicine instrumentation used for production of nuclides, measurement of doses, surveying radioactivity, and imaging (including SPECT, PET, and PET-CT). The final chapters cover radiation biology, radiation safety, and radiation accidents.

Numerous illustrations are included. They are highly schematic and are designed to illustrate concepts rather than represent scale models of their subjects. This text is intended for radiology residents, cardiology fellows, nuclear medicine fellows, nuclear medicine technology students, and others interested in an introduction to concepts in nuclear medicine physics and instrumentation.

RAP

Acknowledgments

The authors would like to thank the following experts for their valuable critiques of portions of this text: Stephen Moore, Ph.D. on the topic of SPECT processing including iterative reconstruction, Fred Fahey, D.Sc. on PET instrumentation, and Robert Zimmerman, M.S.E.E. on gamma camera quality control and the physics of crystal scintillators. In addition, Dr Frank Masse generously reviewed the material on radiation accidents and Mark Walsh, C.H.P. critiqued the radiation safety text. Many thanks to Margaret Nordby for her patient review of the proofs. The authors are grateful to Rhonda M. Powsner, M.D. for her assistance in reviewing the text and proofs.

Since the second edition incorporates the text from the first edition the authors would like to thank the following individuals for their help in reviewing portions of the first edition during it's preparation: David Rockwell, M.D., Maura Dineen-Burton, C.N.M.T., Dipa Patel, M.D., Alfonse Taghian, M.D., Hernan Jara, Ph.D., Susan Gussenhoven, Ph.D., John Shaw, M.S., Michael Squillante, Ph.D., Kevin Buckley, C.H.P., Jayne Caruso, Victor Lee, M.D., Toby Wroblicka, M.D., Dan Winder, M.D., Dennis Atkinson, M.D., and Inna Gazit, M.D.. Thanks to Peter Shomphe, A.R.R.T., C.N.M.T., Bob Dann, Ph.D., and Lara Patriquin, M.D. for wading through the manuscript in its entirety. We greatly appreciate the patience shown at that time by Robert Zimmerman, M.S.E.E., Kevin Buckley, C.H.P., John Widman, Ph.D., C.H.P., Peter Waer, Ph.D., Stephen Moore, Ph.D., Bill Worstell, Ph.D., and Hernan Jara, Ph.D. while answering our numerous questions. Thanks to Delia Edwards, Milda Pitter, and Paul Guidone, M.D. for taking time to pose as models.

RAP

ERP

Contributing author

Kevin Donohoe, M.D.
Staff Physician in Nuclear Medicine
Beth Israel Deaconess Medical Center
Assistant Professor of Radiology
Harvard Medical School

1 Basic nuclear medicine physics

Properties and Structure of Matter

Matter has several fundamental properties. For our purposes the most important are **mass** and **charge** (**electric**). We recognize mass by the force gravity exerts on a material object (commonly referred to as its weight) and by the object's inertia, which is the "resistance" we encounter when we attempt to change the position or motion of a material object.

Similarly, we can, at least at times, recognize charge by the direct effect it can have on us or that we can observe it to have on inanimate objects. For example, we may feel the presence of a strongly charged object when it causes our hair to move or even to stand on end. More often than not, however, we are insensitive to charge. But whether grossly detectable or not, its effects must be considered here because of the role charge plays in the structure of matter.

Charge is generally thought to have been recognized first by the ancient Greeks. They noticed that some kinds of matter, an amber rod for example, can be given an electric charge by rubbing it with a piece of cloth. Their experiments convinced them that there are two kinds of charge: opposite charges, which attract each other, and like charges, which repel. One kind of charge came to be called positive, the other negative. We now know that the negative charge is associated with electrons. The rubbing transferred some of the electrons from the atoms of the matter in the rod to the cloth. In a similar fashion, electrons can be transferred to the shoes of a person walking across a carpet. The carpet will then have a net positive charge and the shoes (and wearer) a net negative charge (Fig. 1-1). With these basic properties in mind, we can look at matter in more detail.

Matter is composed of **molecules**. In any chemically pure material, the molecules are the smallest units that retain the characteristics of the material itself. For example, if a block of salt were to be broken into successively smaller pieces,

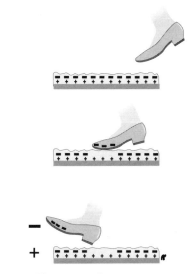

Figure 1-1 Electrostatic charge.

the smallest fragment with the properties of salt would be a single salt molecule (Fig. 1-2). With further fragmentation the molecule would no longer be salt. Molecules, in turn, are composed of **atoms**. Most molecules consist of more than one kind of atom—salt, for example, is made up of atoms of chlorine and sodium. The atoms themselves are composed of smaller particles, the **subatomic particles**, which are discussed later.

The molecule is held together by the chemical bonds among its atoms. These bonds are formed by the force of electrical attraction between oppositely charged parts of the molecule. This force is often referred to as the Coulomb force after Charles A. de Coulomb, the physicist who characterized it. This is the force involved in chemical reactions such as the combining of hydrogen and oxygen to form water. The electrons of the atom

are held by the electrical force between them and the positive nucleus. The nucleus of the atom is held together by another type of force—nuclear force—which is involved in the release of atomic energy. Nuclear forces are of greater magnitudes than electrical forces.

Elements

There are more than 100 species of atoms. These species are referred to as **elements**. Most of the known elements—for example, mercury, helium, gold, hydrogen, and oxygen—occur naturally on earth; others are not usually found in nature but are made by humans—for example, europium and americium. A reasonable explanation for the absence of some elements from nature is that if and when they were formed they proved too unstable to survive in detectable amounts into the present.

All the elements have been assigned symbols or abbreviated chemical names: gold—Au, mercury—Hg, helium—He. Some symbols are obvious abbreviations of the English name; others are derived from the original Latin name of the element, for example, Au is from aurum, the Latin word for gold.

All of the known elements, both natural and those made by humans, are organized in the **periodic table**. In Figure 1-3, the elements that have a

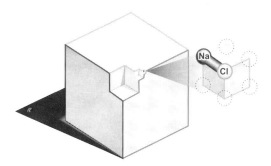

Figure 1-2 The NaCl molecule is the smallest unit of salt that retains the characteristics of salt.

1 H																	2 He
3 Li	4 Be											5 B	6 C	7 N	8 O	9 F	10 Ne
11 Na	12 Mg											13 Al	14 Si	15 P	16 S	17 Cl	18 Ar
19 K	20 Ca	21 Sc	22 Ti	23 V	24 Cr	25 Mn	26 Fe	27 Co	28 Ni	29 Cu	30 Zn	31 Ga	32 Ge	33 As	34 Se	35 Br	36 Kr
37 Rb	38 Sr	39 Y	40 Zr	41 Nb	42 Mo	43 Tc	44 Ru	45 Rh	46 Pd	47 Ag	48 Cd	49 In	50 Sn	51 Sb	52 Te	53 I	54 Xe
55 Cs	56 Ba	57 La	58 Ce	59 Pr	60 Nd	61 Pm	62 Sm	63 Eu	64 Gd	65 Tb	66 Dy	67 Ho	68 Er	69 Tm	70 Yb	71 Lu	
			72 Hf	73 Ta	74 W	75 Re	76 Os	77 Ir	78 Pt	79 Au	80 Hg	81 Tl	82 Pb	83 Bi	84 Po	85 At	86 Rn
87 Fr	88 Ra	89 Ac	90 Th	91 Pa	92 U	93 Np	94 Pu	95 Am	96 Cm	97 Bk	98 Cf	99 Es	100 Fm	101 Md	102 No	103 Lr	
			104 Rf	105 Ha	106 Sg	107 Ns	108 Hs	109 Mt	110 ?	111 ?							

Figure 1-3 Periodic table.

stable state are shown in white boxes; those that occur only in a radioactive form are shown in gray boxes. Elements 104 to 111 have not been formally named (proposed names are listed). When necessary, the chemical symbol shown in the table for each element can be expanded to include three numbers to describe the composition of its nucleus (Fig. 1-4).

Atomic Structure

Atoms initially were thought of as no more than small pieces of matter. Our understanding that they have an inner structure has it roots in the observations of earlier physicists that the atoms of which matter is composed contain **electrons** of negative charge. In as much as the atom as a whole is electrically neutral, it seemed obvious that it must also contain something with a positive charge to balance the negative charge of the electrons. Thus, early attempts to picture the atom, modeled on our solar system, showed the negatively charged electrons orbiting a central group of particles, the positively charged **nucleus** (Fig. 1-5).

Electrons

In our simple solar-system model of the atom, the electrons are viewed as orbiting the nucleus at high speeds. They have a negative charge and are drawn toward the positively charged nucleus. The electrical charges of the atom are "balanced," that is, the total negative charge of the electrons equals the positive charge of the nucleus. As we shall see in a moment, this is simply another way to point out that the number

of orbital electrons equals the number of nuclear protons.

Although each electron orbits at high speed, it remains in its orbit because the electrical force draws it toward the positively charged nucleus. This attraction keeps the moving electron in its orbit in much the same way as a string tied to a ball will hold it in its path as you swing it rapidly around your head (Fig. 1-6).

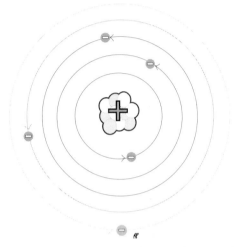

Figure 1-5 Flat atom. The standard two-dimensional drawing of atomic structure.

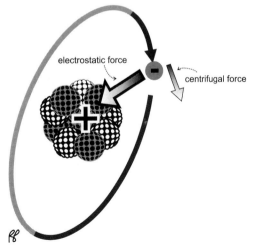

Figure 1-6 The Coulomb force between the negative electrons and the positive protons keeps the electron in orbit. Without this electric force the electron would fly off into space.

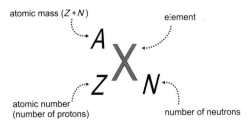

Figure 1-4 Standard atomic notation.

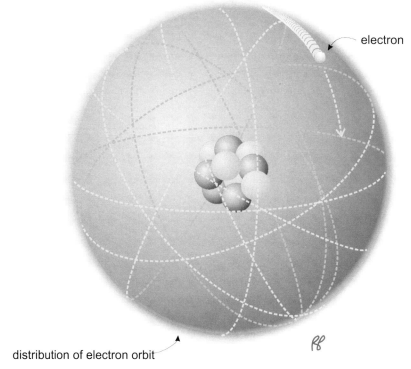

electron

distribution of electron orbit

Figure 1-7 An electron shell is a representation of the energy level in which the electron moves.

Electron Shells

By adding a third dimension to our model of the atom, we can depict the electron orbits as the surfaces of spheres (called **shells**) to suggest that, unlike the planets orbiting the sun, electrons are not confined to a circular orbit lying in a single plane but may be more widely distributed (Fig. 1-7). Of course, neither the simple circular orbits nor these electron shells are physical entities; rather, they are loose representations of the "distances" the orbital electrons are from the nucleus (Fig. 1-8). Although it is convenient for us to talk about distances and diameters of the shells, distance on the atomic scale does not have quite the same meaning it does with everyday objects. The more significant characteristic of a shell is the energy it signifies.

The closer an electron is to the nucleus, the more tightly it is held by the positive charge of nucleus. In saying this, we mean that more work (energy) is required to remove an inner-shell electron than an outer one. The energy that must be put into the atom to separate an electron is called the **electron binding energy**. It is usually expressed in **electron volts (eV)**. The electron binding energy varies from a few thousand electron volts (keV) for inner-shell electrons to just a few eV for the less tightly bound outer-shell electrons.

ELECTRON VOLT

The electron volt is a special unit defined as the energy required to move one electron against a potential difference of one volt. It is a small unit on the everyday scale, at only 1.6×10^{-19} joules (J), but a very convenient unit on the atomic scale. One joule is the Système International (SI) unit of work or energy. For comparison, 1 J equals 0.24 small calories (as opposed to the kcal used to measure food intake).

Figure 1-8 Cut-away model of a medium-sized atom such as argon.

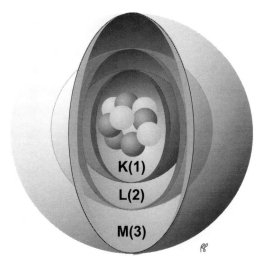

Figure 1-9 K, L, and M electron shells.

Quantum Number

The atomic electrons in their shells are usually described by their **quantum numbers**, of which there are four types. The first is the **principal** quantum number (**n**), which identifies the energy shell. The first three shells (K, L, and M) are depicted in Figure 1-9. The electron binding energy is greatest for the innermost shell (K) and is progressively less for the outer shells. Larger atoms have more shells.

The second quantum number is the **azimuthal** quantum number (**l**), which can be thought of as a **subshell** within the shell. Technically l is the angular momentum of the electron and is related to the product of the mass of the electron, its velocity, and the radius of its orbit. Each subshell is assigned a letter designation: s, p, d, f, and so on. For completeness, the full label of a subshell includes the numeric designation of its principal shell, which for L is the number 2; thus 2s and 2p.

The third number, the **magnetic** quantum number (**m_l**), describes the direction of rotation of the electron and the orientation of the subshell orbit. The fourth quantum number is the **spin** quantum number (**m_s**), which refers to the direction the electron spins on its axis. Both the third and fourth quantum numbers contribute to the **magnetic moment** (or magnetic field) created by the moving electron. The four quantum numbers are outlined in Table 1-1.

QUANTUM NUMBERS

The term quantum means, literally, amount. It acquired its special significance in physics when Bohr and others theorized that physical quantities such as energy and light could not have a range of values as on a continuum, but rather could have only discrete, step-like values. The individual steps are so small that their existence escaped the notice of physicists until Bohr postulated them to explain his theory of the atom. We now refer to Bohr's theory as **quantum theory** and the resulting explanations of motion in the atomic scale as **quantum mechanics** to distinguish it from the classical mechanics described by Isaac Newton, which is still needed for everyday engineering.

The innermost or K shell has only one subshell (called the s subshell). This subshell has a magnetic quantum number of zero and two possible values for the spin quantum number, m_s; these are $+\frac{1}{2}$ and $-\frac{1}{2}$. The neutral atom with a full K shell, that is to say, with two electrons "circling" the nucleus, is the helium atom.

Table 1-1 Quantum Numbers and Values

Quantum Number	Corresponding Names	Range of Values
Principal (**n**)	K, L, M, ...	1, 2, 3, ...
Azimuthal (**l**)	s, p, d, f, g, ...	0, 1, 2, 3, ... (**n** − 1)
Magnetic (**m**$_l$)	None	−**l**, −(**l** − 1), ... 0, ... (**l** − 1), **l**
Spin (**m**$_s$)	Down, up	−$\frac{1}{2}$, +$\frac{1}{2}$

The next shell, the L shell, has available an s subshell and a second subshell (called the 2p subshell). The 2s subshell in the L shell is similar to the s subshell of the K shell and can accommodate two electrons (Fig. 1-10A). The 2p subshell has three possible magnetic quantum numbers (−1, 0, and 1) or subshells, and for each of these quantum numbers there are the two available spin quantum numbers, which allows for a total of six electrons. Each 2p subshell

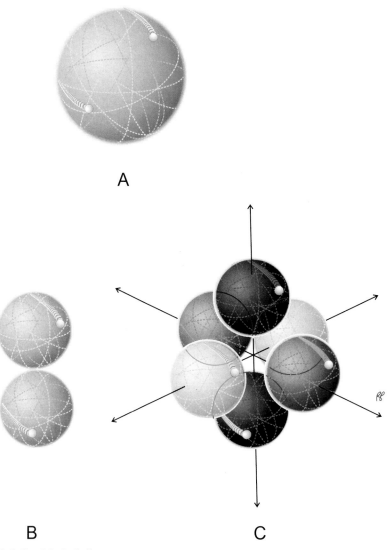

A

B   C

Figure 1-10 Subshells of the L shell.

Table 1-2 Electron Quantum States

Quantum Number Designations	Quantum States									
Principal or radial (n)	1	2								
Azimuthal (l)	0	0	1							
Magnetic (m_l)	0	0	−1		0		1			
Spin (m_s)	$+\frac{1}{2}$	$-\frac{1}{2}$	$+\frac{1}{2}$	$-\frac{1}{2}$	$+\frac{1}{2}$	$-\frac{1}{2}$	$+\frac{1}{2}$	$-\frac{1}{2}$	$+\frac{1}{2}$	$-\frac{1}{2}$

Element		Number of Electrons in Each State									
Atomic Number	Chemical Name										
1	Hydrogen	1	0	0	0	0	0	0	0	0	0
2	Helium	1	1	0	0	0	0	0	0	0	0
3	Lithium	1	1	1	0	0	0	0	0	0	0
4	Beryllium	1	1	1	1	0	0	0	0	0	0
5	Boron	1	1	1	1	1	0	0	0	0	0
6	Carbon	1	1	1	1	1	1	0	0	0	0
7	Nitrogen	1	1	1	1	1	1	1	0	0	0
8	Oxygen	1	1	1	1	1	1	1	1	0	0
9	Flourine	1	1	1	1	1	1	1	1	1	0
10	Neon	1	1	1	1	1	1	1	1	1	1

is depicted as two adjacent spheres, a kind of three-dimensional figure eight (Fig. 1-10B). The arrangement of all three subshells is shown in Figure 1-10C. The L shell can accommodate a total of eight electrons. The neutral atom containing all ten electrons in the K and L shells is neon.

The number of electrons in each set of shells for the light elements, hydrogen through neon, forms a regular progression, as shown in Table 1-2. For the third and subsequent shells, the ordering and filling of the subshells, as dictated by the rules of quantum mechanics, is less regular and will not be covered here.

Stable Electron Configuration

Just as it takes energy to remove an electron from its atom, it takes energy to move an electron from an inner shell to an outer shell, which can also be thought of as the energy required to pull a negative electron away from the positively charged nucleus. Any vacancy in an inner shell creates an unstable condition often referred to as an **excited state**.

The electrical charges of the atom are balanced, that is, the total negative charge of the electrons equals the total positive charge of the nucleus. This is simply another way of pointing out that the number of orbital electrons equals the number of nuclear protons. Furthermore, the electrons must fill the shells with the highest binding energy first. At least in the elements of low atomic number, electrons in the inner shells have the highest binding energy.

If the arrangement of the electrons in the shells is not in the stable state, they will undergo rearrangement in order to become stable, a process often referred to as **de-excitation**. Because the stable configuration of the shells always has less energy than any unstable configuration, the de-excitation releases energy as photons, often as **x-rays**.

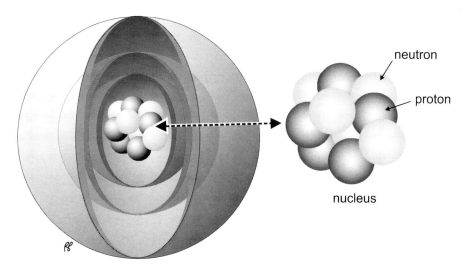

Figure 1-11 The nucleus of an atom is composed of protons and neutrons.

Table 1-3 The Subatomic Particles

Name	Symbol	Location	Mass[a]	Charge
Neutron	N	Nucleus	1840	None
Proton	P	Nucleus	1836	Positive (+)
Electron	e⁻	Shell	1	Negative (−)

[a] Relative to an electron.

Nucleus

Like the atom itself, the atomic nucleus also has an inner structure (Fig. 1-11). Experiments showed that the nucleus consists of two types of particles: **protons**, which carry a positive charge, and **neutrons**, which carry no charge. The general term for protons and neutrons is **nucleons**. The nucleons, as shown in Table 1-3, have a much greater mass than electrons. Like electrons, nucleons have quantum properties including spin. The nucleus has a spin value equal to the sum of the nucleon spin values.

A simple but useful model of the nucleus is a tightly bound cluster of protons and neutrons. Protons naturally repel each other since they are positively charged; however, there is a powerful binding force called the **nuclear force** that holds the nucleons together very tightly

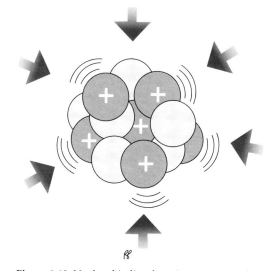

Figure 1-12 Nuclear binding force is strong enough to overcome the electrical repulsion between the positively charged protons.

(Fig. 1-12). The work (energy) required to overcome the nuclear force, the work to remove a nucleon from the nucleus, is called the **nuclear binding energy**. Typical binding energies are in the range of 6 million to 9 million electron volts (MeV) (approximately one thousand to one million times the electron binding force). The magnitude of the binding energy is related to

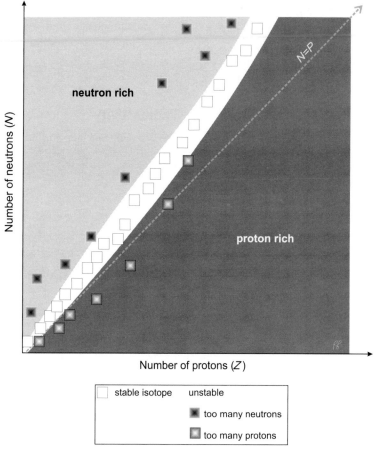

Figure 1-13 All combinations of neutrons and protons that can coexist in a stable nuclear configuration lie within the broad white band.

another fact of nature: the measured mass of a nucleus is always less than the mass expected from the sum of the masses of its neutrons and protons. The "missing" mass is called the **mass defect**, the energy equivalent of which is equal to the nuclear binding energy. This interchangeability of mass and energy was immortalized in Einstein's equation $E = mc^2$.

The Stable Nucleus

Not all elements have stable nuclei; they do exist for most of the light and mid-weight elements, those with atomic numbers up to and including bismuth ($Z = 83$). The exceptions are technetium ($Z = 43$) and promethium ($Z = 61$). All those with atomic numbers higher than 83, such as

radium ($Z = 88$) and uranium ($Z = 92$), are inherently unstable because of their large size.

For those nuclei with a stable state there is an optimal ratio of neutrons to protons. For the lighter elements this ratio is approximately 1 : 1; for increasing atomic weights, the number of neutrons exceeds the number of protons. A plot depicting the number of neutrons as a function of the number of protons is called the **line of stability**, depicted as a broad white band in Figure 1-13.

Isotopes, Isotones, and Isobars

Each atom of any sample of an element has the same number of protons (the same **Z**: atomic number) in its nucleus. Lead found anywhere in the world will always be composed of atoms with 82 protons. The same does not apply, however, to the number of neutrons in the nucleus.

An **isotope** of an element is a particular variation of the nuclear composition of the atoms of that element. The number of protons (Z: atomic number) is unchanged, but the number of neutrons (N) varies. Since the number of neutrons changes, the total number of neutrons and protons (A: the atomic mass) changes. Two related entities are **isotones** and **isobars**. Isotones

are atoms of different elements that contain identical numbers of neutrons but varying numbers of protons. Isobars are atoms of different elements with identical numbers of nucleons. Examples of these are illustrated in Figure 1-14.

Radioactivity

The Unstable Nucleus and Radioactive Decay

A nucleus not in its stable state will adjust itself until it is stable either by ejecting

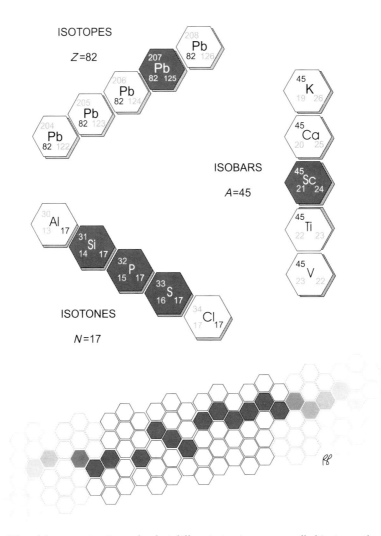

Figure 1-14 Nuclides of the same atomic number but different atomic mass are called isotopes, those of an equal number of neutrons are called isotones, and those of the same atomic mass but different atomic number are called isobars.

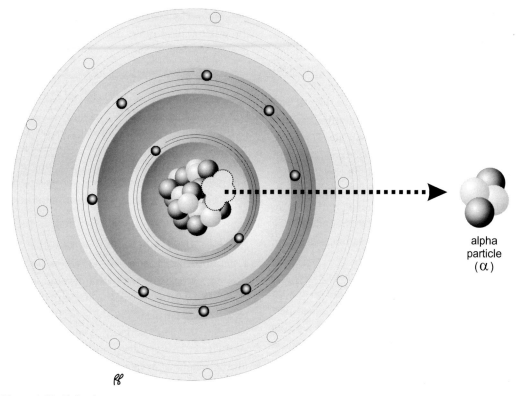

Figure 1-15 Alpha decay.

portions of its nucleus or by emitting energy in the form of photons (**gamma rays**). This process is referred to as **radioactive decay**. The type of decay depends on which of the following rules for nuclear stability is violated.

Excessive Nuclear Mass

Alpha Decay

Very large unstable atoms, atoms with high atomic mass, may split into nuclear fragments. The smallest stable nuclear fragment that is emitted is the particle consisting of two neutrons and two protons, equivalent to the nucleus of a helium atom. Because it was one of the first types of radiation discovered, the emission of a helium nucleus is called **alpha radiation**, and the emitted helium nucleus is called an **alpha particle** (Fig. 1-15).

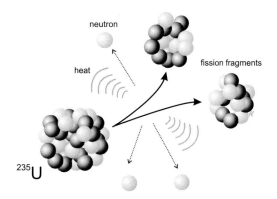

Figure 1-16 Fission of a ^{235}U nucleus.

Fission

Under some circumstances, the nucleus of the unstable atom may break into larger fragments, a process usually referred to as **nuclear fission**. During fission two or three neutrons and heat are emitted (Fig. 1-16).

Unstable Neutron–Proton Ratio

Too Many Neutrons: Beta Decay

Nuclei with excess neutrons can achieve stability by a process that amounts to the conversion of a neutron into a proton and an electron. The proton remains in the nucleus, but the electron is emitted. This is called **beta radiation**, and the electron itself is called a **beta particle** (Fig. 1-17). The process and the emitted electron were given these names to contrast with the alpha particle before the physical nature of either was discovered. The beta particle generated in this decay will become a free electron until it finds a vacancy in an electron shell either in the atom of its origin or in another atom.

Careful study of beta decay suggested to physicists that the conversion of neutron to proton involved more than the emission of a beta particle (electron). Beta emission satisfied the rule for conservation of charge in that the neutral neutron yielded one positive proton and one negative electron; however, it did not appear to satisfy the equally important rule for conservation of energy. Measurements showed that most of the emitted electrons simply did not have all the energy expected. To explain this apparent discrepancy, the emission of a second particle was postulated and that particle was later identified experimentally. Called an **antineutrino** (neutrino for small and neutral), it carries the "missing" energy of the reaction.

Too Many Protons: Positron Decay and Electron Capture

In a manner analogous to that for excess neutrons, an unstable nucleus with too many protons can undergo a decay that has the effect of converting a proton into a neutron. There are two ways this can occur: positron decay and electron capture.

Positron decay: A proton can be converted into a neutron and a **positron**, which is an electron with a positive, instead of negative, charge (Fig. 1-18). The positron is also referred to as a positive beta particle or positive electron or anti-electron. In positron decay, a **neutrino** is also emitted. In many ways, positron decay is the mirror image of beta decay: positive electron instead of negative electron, neutrino instead of antineutrino. Unlike the negative electron, the positron itself survives only briefly. It quickly encounters an electron (electrons are plentiful in matter), and both are **annihilated** (see Fig. 8-1). This is why it is considered an anti-electron.

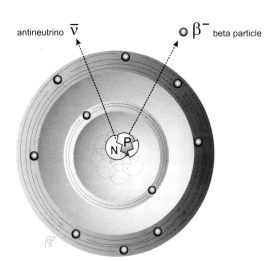

Figure 1-17 β^- (negatron) decay.

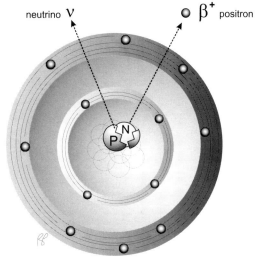

Figure 1-18 β^+ (positron) decay.

Generally speaking, antiparticles react with the corresponding particle to annihilate both.

During the annihilation reaction, the combined mass of the positron and electron is converted into two photons of energy equivalent to the mass destroyed. Unless the difference between the masses of the parent and daughter atoms is at least equal to the mass of one electron plus one positron, a total equivalent to 1.02 MeV, there will be insufficient energy available for positron emission.

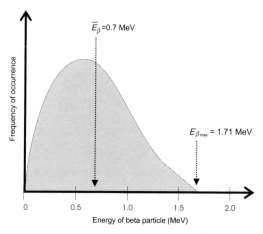

Figure 1-19 Beta emissions (both β^- and β^+) are ejected from the nucleus with energies between zero and their maximum possible energy ($E_{\beta\,max}$). The average energy ($\bar{E}_\beta$) is equal to approximately one third of the maximum energy. This is an illustration of the spectrum of emissions for 32P.

ENERGY OF BETA PARTICLES AND POSITRONS

Although the total energy emitted from an atom during beta decay or positron emission is constant, the relative distribution of this energy between the beta particle and antineutrino (or positron and neutrino) is variable. For example, the total amount of available energy released during beta decay of a phosphorus-32 atom is 1.7 MeV. This energy can be distributed as 0.5 MeV to the beta particle and 1.2 MeV to the antineutrino, or 1.5 MeV to the beta particle and 0.2 MeV to the antineutrino, or 1.7 MeV to the beta particle and no energy to the antineutrino, and so on. In any group of atoms the likelihood of occurrence of each of such combinations is not equal. It is very uncommon, for example, that all of the energy is carried off by the beta particle. It is much more common for the particle to receive less than half of the total amount of energy emitted. This is illustrated by Figure 1-19, a plot of the number of beta particles emitted at each energy from zero to the maximum energy released in the decay. $E_{\beta max}$ is the maximum possible energy that a beta particle can receive during beta decay of any atom, and $\bar{E}_\beta$ is the average energy of all beta particles for decay of a group of such atoms. The average energy is approximately one-third of the maximum energy

$$\bar{E}_\beta \cong \tfrac{1}{3} E_{\beta max} \qquad\qquad (Eq.\ 1\text{-}1)$$

Electron capture: Through a process that competes with positron decay, a nucleus can combine with one of its inner orbital electrons to achieve the net effect of converting one of the protons in the nucleus into a neutron (Fig. 1-20). An outer-shell electron then fills the vacancy in the inner shell left by the captured electron. The energy lost by the "fall" of the outer-shell electron to the inner shell is emitted as an x-ray.

Appropriate Numbers of Nucleons, but Too Much Energy

If the number of nucleons and the ratio of neutrons to protons are both within their stable ranges, but the energy of the nucleus is greater than its resting level (an excited state), the excess energy is shed by **isomeric transition**. This may occur by either of the competing reactions, gamma emission or internal conversion.

Gamma Emission

In this process, excess nuclear energy is emitted as a **gamma ray** (Fig. 1-21). The name gamma was given to this radiation, before its physical nature was understood, because it was the third (alpha, beta, gamma) type of radiation discovered. A gamma ray is a photon (energy) emitted by an excited nucleus. Despite its unique name, it cannot be distinguished from photons of the

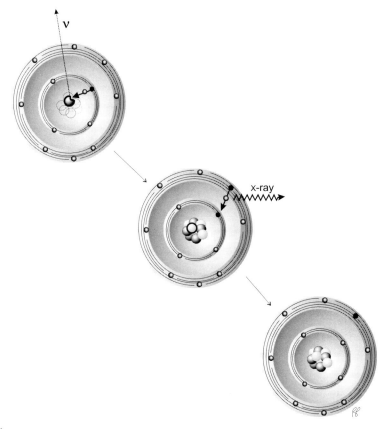

Figure 1-20 Electron capture.

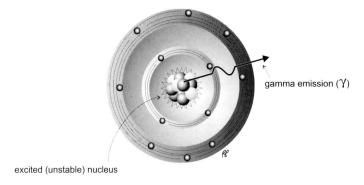

excited (unstable) nucleus

Figure 1-21 Isomeric transition. Excess nuclear energy is carried off as a gamma ray.

same energy from different sources, for example x-rays.

Internal Conversion

The excited nucleus can transfer its excess energy to an orbital electron (generally an inner-shell electron) causing the electron to be ejected from the atom. This can only occur if the excess energy is greater than the binding energy of the electron. This electron is called a **conversion electron** (Fig. 1-22). The resulting inner orbital vacancy is rapidly filled with an outer-shell electron (as the atom assumes a more stable state, inner orbitals are filled before outer orbitals).

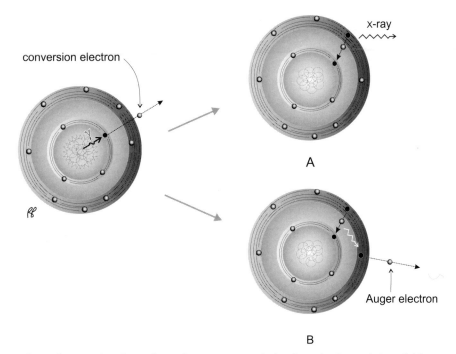

x-ray

conversion electron

A

Auger electron

B

Figure 1-22 Internal conversion. As an alternative to gamma emission, it can lead to emission of either an x-ray (A) or an Auger electron (B).

The energy released as a result of the "fall" of an outer-shell electron to an inner shell is emitted as an x-ray or as a free electron (**Auger electron**).

Table 1-4 reviews the properties of the various subatomic particles.

Decay Notation

Decay of a nuclide from an unstable (excited) to a stable (ground) state can occur in a series of steps, with the production of particles and photons characteristic of each step. A standard notation is used to describe these steps (Fig. 1-23). The uppermost level of the schematic is the state with the greatest energy. As the nuclide decays by losing energy and/or particles, lower horizontal levels represent states of relatively lower energy. Directional arrows from one level to the next indicate the type of decay. By convention, an oblique line angled downward and to the left indicates electron capture; downward and to the right, beta emission; and a vertical arrow, an isomeric

transition. The dogleg is used for positron emission. Notice that a pathway ending to the left, as in electron capture or positron emission, corresponds to a decrease in atomic number. On the other hand, a line ending to the right, as in beta emission, corresponds to an increase in atomic number.

Figure 1-24 depicts specific decay schemes for ^{99m}Tc, ^{111}In, and ^{131}I. The "m" in ^{99m}Tc stands for **metastable**, which refers to an excited nucleus with an appreciable lifetime ($>10^{-12}$ seconds) prior to undergoing isomeric transition.

Half-Life

It is not possible to predict when an individual nuclide atom will decay, just as in preparing popcorn one cannot determine when any particular kernel of corn will open. However, the average behavior of a large number of the popcorn kernels is predictable. From experience with microwave popcorn, one knows that half of the kernels will pop within 2 min and most of the

Table 1-4 Properties of the Subatomic Particles

Name(s)	Symbol	Mass[a]	Charge
Neutron	N	1840	None
Proton	P	1836	Positive (+)
Electron	e^-	1	Negative (−)
Beta particle (beta minus particle, electron)[b]	β^-	1	Negative (−)
Positron (beta plus particle, positive electron)	β^+	1	Positive (+)
Gamma ray (photon)	γ	None	None
X-ray	X-ray	None	None
Neutrino	ν	Near zero	None
Antineutrino	$\bar{\nu}$	Near zero	None

[a] Relative to an electron.
[b] There is no physical difference between a beta particle and an electron; the term beta particle is applied to an electron that is emitted from a radioactive nucleus. The symbol β without a minus or plus sign attached always refers to a beta minus particle or electron.

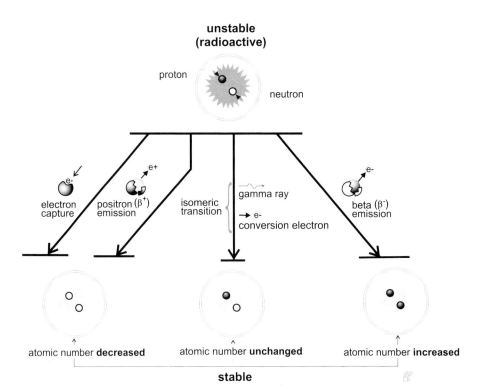

Figure 1-23 Decay schematics.

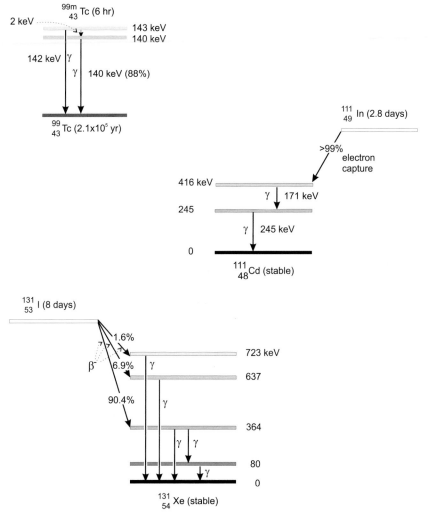

Figure 1-24 Decay schemes showing principal transitions for technetium-99m, indium-111, iodine-131. Energy levels are rounded to three significant figures.

bag will be done in 4 min. In a like manner, the average behavior of a radioactive sample containing billions of atoms is predictable. The time it takes for half of these atoms to decay is called (appropriately enough) the **half-life**, or in scientific notation $T_{1/2}$ (pronounced "T one-half"). It is not surprising that the time it takes for half of the remaining atoms to decay is also $T_{1/2}$. This process continues until the number of nuclide atoms eventually comes so close to zero that we can consider the process complete. A plot of $A(t)$, the activity remaining, is shown in Figure 1-25. This

curve, and therefore the average behavior of the sample of radioactivity, can be described by the **decay equation**:

$$A(t) = A(0)e^{-0.693t/T_{1/2}} \qquad (Eq.\ 1\text{-}2)$$

where $A(0)$ is the initial number of radioactive atoms.

A commonly used alternative form of the decay equation employs the **decay constant** (λ), which is approximately 0.693 divided by the half-life ($T_{1/2}$):

$$\lambda = 0.693/T_{1/2} \qquad (Eq.\ 1\text{-}3)$$

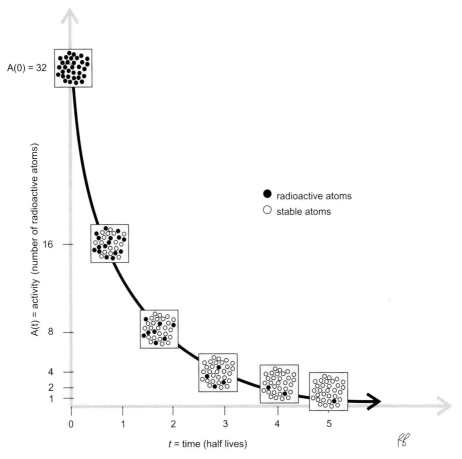

Figure 1-25 Decay curve. Note the progressive replacement of radioactive atoms by stable atoms as shown schematically in each box.

The decay equation can be rewritten as

$$A(t) = A(0)e^{-\lambda t} \qquad (Eq.\ 1\text{-}4)$$

The amount of activity of any radionuclide may be expressed as the number of decays per unit time. Common units for measuring radioactivity are the **curie** (after Marie Curie) or the newer SI unit, the **becquerel** (after another nuclear pioneer, Henri Becquerel). One becquerel is defined as one radioactive decay per second. Nuclear medicine doses are generally a million times greater and are more easily expressed in megabecquerels (MBq). One curie (Ci) is defined as 3.7×10^{10} decays per second (this was picked because it is approximately equal

to the radioactivity emitted by 1 g of radium in equilibrium with its daughter nuclides). A partial list of conversion values is provided in Table 1-5.

A related term that is frequently confused with decay is the **count**, which refers to the registration of a single decay by a detector such as a Geiger counter. Most of the detectors used in nuclear medicine detect only a fraction of the decays, principally because the radiation from many of the decays is directed away from the detector. Count rate refers to the number of decays actually counted in a given time, usually counts per minute. All things being equal, the count rate will be proportional to the decay rate, and

Table 1-5 Conversion Values for Units of Radioactivity

One curie (Ci)	One millicurie (mCi)	One microcurie (μCi)	One bequerel (Bq)[a]	One megabecquerel (MBq)
	10^{-3} Ci	10^{-6} Ci	27×10^{-12} Ci	27×10^{-6} Ci
1×10^3 mCi		10^{-3} mCi	27×10^{-9} mCi	27×10^{-3} mCi
1×10^6 μCi	1×10^3 μCi		27×10^{-6} μCi	27 μCi
37×10^9 Bq	37×10^6 Bq	37×10^3 Bq		1×10^6 Bq
37×10^3 MBq	37 MBq	37×10^{-3} MBq	1×10^{-6} MBq	

[a] One becquerel equals one decay per second.

it is a commonly used, if inexact, measure of radioactivity.

Test Yourself

1 For each of the five terms below, choose the best definition
(1) Isobars
(2) Isoclines
(3) Isomers
(4) Isotones
(5) Isotopes
   (a) Atoms of the same element (equal Z) with different numbers of neutrons (N)
   (b) Atoms of different elements (different Z) with equal numbers of neutrons (N)
   (c) Atoms of different elements with equal atomic mass (A).
   (d) None of the above, usually used as a geological term.
   (e) Atoms of equal atomic mass (A) and equal atomic number (Z), but with unstable nuclei which exist in different energy states.

2 Which of the following statements are correct?
   (a) There is a stable isotope of technetium.
   (b) Atoms with atomic numbers (Z) > 83 are inherently unstable.

(c) For light elements nuclear stability is achieved with equal numbers of protons and neutrons; for heavier elements the number of neutrons exceeds the number of protons.

3 For internal conversion to occur, the excess energy of the excited nucleus must equal or exceed:
   (a) 0.551 eV
   (b) 1.102 eV
   (c) the internal conversion coefficient
   (d) the average energy of the Auger electrons
   (e) the binding energy of the emitted electron.

4 For an atom undergoing beta decay, the average energy of the emitted beta particles is approximately:
   (a) 0.551 eV
   (b) 0.551 times the loss of atomic mass
   (c) one half of the total energy released for the individual event
   (d) one third of the maximum energy of the emitted beta particles
   (e) equal to the average energy of the accompanying antineutrinos.

5 You receive a dose of ^{99m}Tc measuring 370 MBq from the radiopharmacy at 10 AM. Your patient does not arrive in the department until 2 PM. How much activity, in millicurie, remains? (The $T_{1/2}$ of ^{99m}Tc is 6 hours. $e = 2.718$).

CHAPTER 2

2 Interaction of radiation with matter

When radiation strikes matter, both the nature of the radiation and the composition of the matter affect what happens. The process begins with the transfer of radiation energy to the atoms and molecules, heating the matter or even modifying its structure.

If all the energy of a bombarding particle or photon is transferred, the radiation will appear to have been stopped within the irradiated matter. Conversely, if the energy is not completely deposited in the matter, the remaining energy will emerge as though the matter were transparent or at least translucent. This said, we will now introduce some of the physical phenomena involved as radiation interacts with matter, and in particular we shall consider, separately at first, the interactions in matter of both photons (gamma rays and x-rays) and charged particles (alpha and beta particles).

Interaction of Photons with Matter

As they pass through matter, photons interact with atoms. The type of interaction is a function of the energy of the photons and the atomic number (Z) of elements composing the matter.

Types of Photon Interactions in Matter

In the practice of nuclear medicine, where gamma rays with energies between 50 keV and 550 keV are used, **Compton scattering** is the dominant type of interaction in materials with lower atomic numbers, such as human tissue ($Z = 7.5$). The **photoelectric effect** is the dominant type of interaction in materials with higher atomic numbers, such as lead ($Z = 82$). A third type of interaction of photons with matter, **pair production**, only occurs with very high photon energies (greater than 1020 keV) and is therefore not important in clinical nuclear medicine. Figure 2-1 depicts the predominant type of interaction for various combinations of incident photons and absorber atomic numbers.

Compton Scattering

In Compton scattering the incident photon transfers part of its energy to an outer shell or (essentially) "free" electron, ejecting it from the atom. Upon ejection this electron is called a **Compton electron**. The photon is scattered (Fig. 2-2) at an angle that depends on the amount of energy transferred from the photon to the electron. The scattering angle can range from nearly 0° to 180°. Figure 2-3 illustrates scattering angles of 135° and 45°.

Photoelectric Effect

A gamma ray of low energy, or one that has lost most of its energy through Compton interactions, may transfer its remaining energy to an orbital (generally inner-shell) electron. This process is

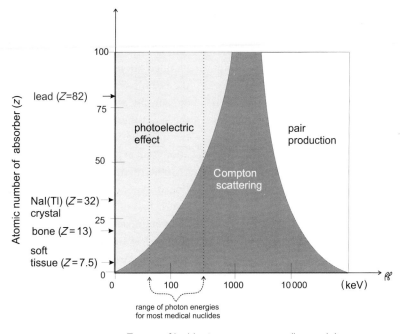

Figure 2-1 Predominant type of interaction for various combinations of incident photons and absorber atomic numbers.

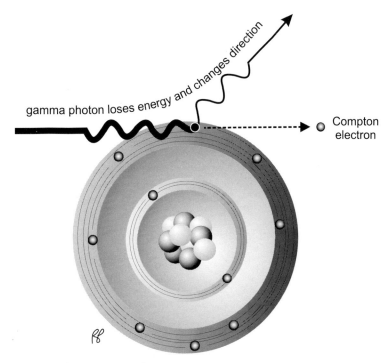

Figure 2-2 Compton scattering.

called the **photoelectric effect** and the ejected electron is called a **photoelectron** (Fig. 2-4). This electron leaves the atom with an energy equal to the energy of the incident gamma ray diminished by the binding energy of the electron. An outer-shell electron then fills the inner-shell vacancy and the excess energy is emitted as an x-ray.

$$E_{photoelectron} = E_{photon} - E_{binding} \quad (Eq. \ 2\text{-}1)$$

Table 2-1 lists the predominant photon interactions in some common materials.

Attenuation of Photons in Matter

As the result of the interactions between photons and matter, the **intensity** of the **beam** (stream of

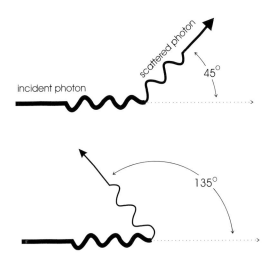

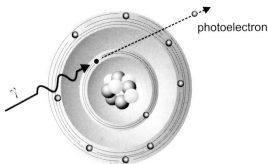

Figure 2-3 Angle of photon scattering.

photons), that is, the number of photons remaining in the beam, decreases as the beam passes through matter (Fig. 2-5). This loss of photons is called **attenuation**; the matter through which the beam passes is referred to as the attenuator. Specifically, attenuation is the ratio of intensity at the point the beam exits the attenuator, I_{out}, to the intensity it had when it entered, I_{in}. Attenuation is an exponential function of the thickness, x, of the attenuator in centimeters. That the function is exponential can be understood to mean that if half of the beam is lost in traversing the first centimeter of material, half of the remainder will be lost traversing the next centimeter, and so on. This resembles the exponential manner in which radioactivity decays with time. Expressed symbolically,

$$I_{out}/I_{in} = e^{-(\mu x)} \quad (Eq. \ 2\text{-}2)$$

where μ, the **linear attenuation coefficient**, is a property of the attenuator. When, as is usually the case, thickness is given in centimeters, the linear attenuation coefficient is expressed as "per centimeter." As might be expected, the linear attenuation coefficient is greater for dense tissue such as bone than for soft tissue such as fat. In general, the linear attenuation coefficient depends on both the energy of the photons and on the average atomic number (Z) and thickness of the attenuator. The lower the energy of the photons or the greater the average atomic number or thickness of the attenuator, the greater the attenuation (Fig. 2-6).

Figure 2-4 Photoelectric effect.

A separate term, the **mass attenuation coefficient** (μ/ρ), is the linear attenuation coefficient divided by the density of the attenuator. When the density of a material is given in grams/cm^3 the units of the mass attenuation coefficient are cm^2/gram.

Absorption of radiation describes another aspect of the process of attenuation. Attenuation describes the weakening of the beam as it passes through matter. Absorption describes the transfer of energy from the beam to the matter.

Table 2-1 Predominant Photon Interactions in Common Materials

Material	Atomic Number (Z)	Density (gm/cc)	Predominant Interaction
H$_2$O	7.4	1.0	Compton scatter
Soft tissue	7.5	1.0	Compton scatter
Glass (silicon)	14	2.6	Compton scatter
O$_2$(gas)	16	0.0014	Compton scatter
NaI (crystal)	32	3.7	Photoelectric effect
Lead	82	11.3	Photoelectric effect
Leaded glass	14, 82	4.8–6.2	Photoelectric effect

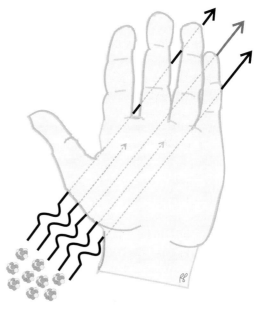

Figure 2-5 Attenuation.

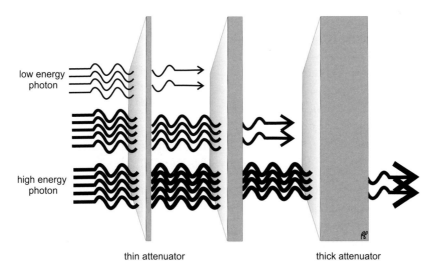

low energy photon

high energy photon

thin attenuator

thick attenuator

Figure 2-6 Half-value layer.

Table 2-2 HVL, TVL, and μ of Lead for Photons of Common Medical Nuclides

Nuclide	Gamma Energy (keV)	Half-Value Layer (cm)	Tenth-Value Layer (cm) (3.32 × HVL)	Linear Attenuation Coefficient, μ (cm^{-1})
^{99m}Tc	140	0.03	0.10	23.10
^{67}Ga	89–389	0.10	0.33	6.93
^{123}I	156	0.04	0.13	17.30
^{131}I	364	0.30	1.00	2.31

Half-Value and Tenth-Value Layers

A material's effectiveness as a photon attenuator is described by the attenuation coefficient. An alternative descriptor, one that is more easily visualized, is the "**half-value layer**" (HVL), which is simply the thickness of a slab of the attenuator that will remove exactly one half of the radiation of a beam. A second slab of the same thickness will remove half of the remainder, leaving one quarter of the original beam, and so forth. For a gamma photon of 100 keV, the HVL in soft tissue is about 4 cm [1].

For any attenuator the **HVL** can be determined experimentally using a photon source and a suitable detector. For calculations involving attenuation of high-intensity radiation beams, an entirely similar concept, the **tenth-value layer** (TVL), is useful. The **TVL** is the thickness of the attenuator that will transmit only one-tenth of the photons in the beam. Two such thicknesses will transmit only one-hundredth of the beam. Table 2-2 lists half- and tenth-value layer as well as the linear attenuation coefficient, μ, of lead for photons of some common medical nuclides.

The linear attenuation coefficient, μ, introduced above, can be calculated from the HVL as follows:

$$\mu = 0.693/\text{HVL} \qquad (Eq.\ 2\text{-}3)$$

The term **penetrating radiation** may be used to describe x-ray and gamma radiation, as they have the potential to penetrate a considerable thickness of any material. Although we have just described some of the many ways photons interact with matter, the likelihood of any of these

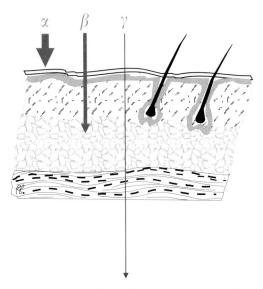

Figure 2-7 Penetrating radiation and nonpenetrating radiation.

interactions occurring over a short distance is small. An individual photon may travel several centimeters or farther into tissue before it interacts. In contrast, charged particles (alpha, beta) undergo many closely spaced interactions. This sharply limits their penetration (Fig. 2-7).

Interaction of Charged Particles with Matter

Because of the strong electrical force between a charged particle and the atoms of an absorber, charged particles can be stopped by matter with relative ease. Compared to photons, they transfer a greater amount of energy in a shorter distance and come to rest more rapidly. For this reason they are referred to as **nonpenetrating radiation** (see Fig. 2-7). In contrast to a photon of 100 keV, an electron of this energy would penetrate less than 0.00014 cm in soft tissue [1] .

Excitation

Charged particles (alphas, betas, and positrons) interact with the electrons surrounding the atom's nucleus by transferring some of their kinetic energy to the electrons. The energy transferred from a low-energy particle is often only sufficient to bump an electron from an inner to an outer shell of the atom. This process is called **excitation**. Following excitation, the displaced electron promptly returns to the lower-energy shell, releasing its recently acquired energy as an x-ray in a process called de-excitation (Fig. 2-8). Because the acquired energy is equal to the difference in binding energies of the electron shells and the binding energies of the electron shells are determined by the atomic structure of the element, the x-ray is referred to as a **characteristic x-ray**.

Ionization

Charged particles of sufficient energy may also transfer enough energy to an electron (generally one in an outer shell) to eject the electron from the atom. This process is called **ionization** (Fig. 2-9). This hole in the outer shell is rapidly filled with an unbound electron. If an inner shell electron is ionized (a much less frequent occurrence) an outer shell electron will "drop" into the inner shell hole and a characteristic x-ray will be emitted. Ionization is not limited to the interaction of charged particles and matter. The photoelectric effect and Compton interactions are examples of photon interactions with matter that produce ionization.

Specific Ionization

When radiation causes the ejection of an electron from an atom of the absorber, the resulting positively charged atom and free negatively

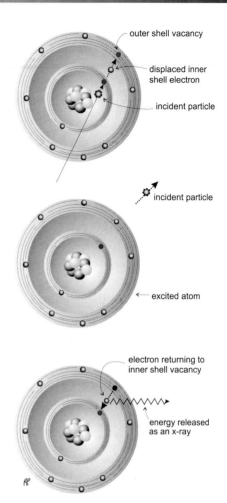

Figure 2-8 Excitation and de-excitation.

charged electron are called an **ion pair** (Fig. 2-9). The amount of energy transferred per ion pair created, *W*, is characteristic of the materials in the absorber. For example, approximately 33 eV (range 25 eV to 40 eV) is transferred to the absorber for each ion pair created in air or water. It is often convenient to refer to the number of ion pairs created per unit distance the radiation travels as its **specific ionization** (*SI*).

Particles with more charge (alpha particles) have a higher specific ionization than lighter particles (electrons).

Linear Energy Transfer

Linear energy transfer (*LET*) is the amount of energy transferred in a given distance by

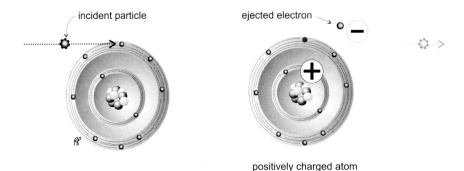

Figure 2-9 Ionization.

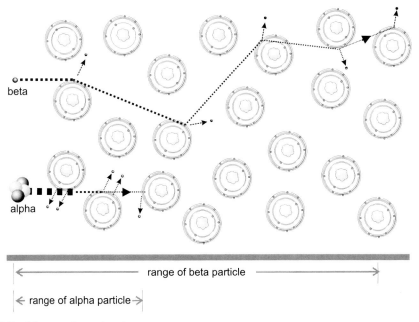

Figure 2-10 Particle range in an absorber.

a particle moving through an absorber. Linear energy transfer is related to specific ionization.

$$LET = SI \times W \qquad (Eq. 2\text{-}4)$$

Alpha particles are classified as high LET radiation, beta particles and photons as low LET radiation.

Range
Range is the distance radiation travels through the absorber. Particles that are lighter, have less charge (such as beta particles), and/or have greater energy travel farther than particles that are heavier, have a greater charge (such as alpha particles), and/or have less energy (Fig. 2-10).

In traversing an absorber, an electron loses energy at each interaction with the atoms of the absorber. The energy loss per interaction is variable. Therefore, the total distance traveled by electrons of the same energy can vary by as much as 3% to 4%. This variation in range is

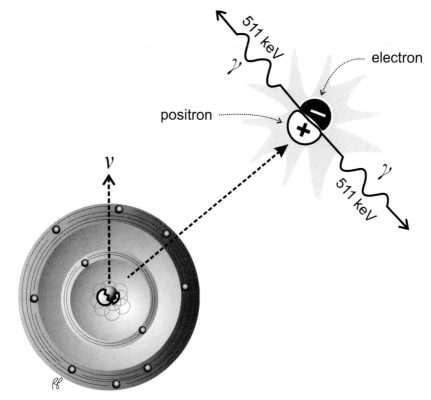

Figure 2-11 Annihilation reaction.

called the **straggling of the ranges**. The heavier alpha particles are not affected to a significant degree and demonstrate very little straggling of range.

Annihilation

This interaction in matter most often involves a positron (positive electron) and an electron (negatron). After a positron has transferred most of its kinetic energy by ionization and excitation, it combines with a free or loosely bound negative electron. Recall that electrons and positrons have equal mass but opposite electric charge. This interaction is explosive, as the combined mass of the two particles is instantly converted to energy in the form of two oppositely directed photons, each of 511 keV. This is referred to as an **annihilation reaction**

$$E = mc^2$$

(energy = mass times the speed of light squared)

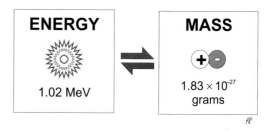

Figure 2-12 Einstein's theory of the equivalence of energy and mass.

(Fig. 2-11). It is another example of the interchangeability of mass and energy described in Einstein's equation: energy equals mass times the speed of light squared, or $E = mc^2$ (Fig. 2-12).

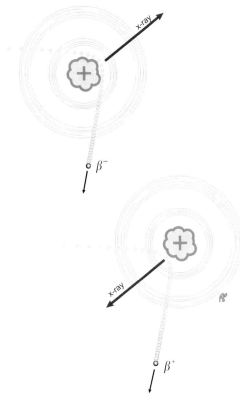

Figure 2-13 Bremsstrahlung. Beta particles (β^-) and positrons (β^+) that travel near the nucleus will be attracted or repelled by the positive charge of the nucleus, generating x-rays in the process.

Bremsstrahlung

Small charged particles such as electrons or positrons may be deflected by nuclei as they pass through matter, which may be attributed to the positive charge of the atomic nuclei. This type of interaction generates x-radiation known as **bremsstrahlung** (Fig. 2-13), which in German means "braking radiation."

Reference

1 Shapiro, J. *Radiation Protection. A Guide for Scientists, Regulators, and Physicians*, 4th Edition, Harvard University Press, Cambridge MA, 2002, pp. 42 and 53.

Test Yourself

1 Which of the following is true of the interaction of charged particles with matter?
 (a) Alpha particles have a higher LET than beta particles.
 (b) The range of alpha particles is generally greater than beta particles.
 (c) Alpha particles have a higher specific ionization than beta particles.

2 True or false: Bremsstrahlung is x-ray radiation emitted as moving electron or positron slows down and is deflected in close proximity to a nucleus.

3 True or false: The photoelectric effect is the dominant type of photon interaction in tissue for radionuclides used in the practice of nuclear medicine.

4 For each of the terms listed here, select the appropriate definition.
 (a) HVL (half-value layer)
 (b) TVL (tenth-value layer)
 (c) μ (linear attenuation coefficient)
 (d) θ (absorption coefficient)
   (i) Thickness of an attenuator that will reduce the intensity (number of photons) in a beam by 90%
   (ii) Thickness of an attenuator that will reduce the intensity (number of photons) in a beam by 50%
   (iii) 0.693/HVL.

5 Which of the following occur during photon interactions with matter and which occur during charged particle interactions with atoms (more than one answer may apply)?
 (a) Excitation
 (b) Pair production
 (c) Ionization
 (d) Compton scattering
 (e) Bremsstrahlung
 (f) Photoelectric effect
 (g) Annihilation reaction.

3

Formation of radionuclides

Many radionuclides exist in nature. An example is ^{14}C, which decays slowly with a half-life of 5700 years and is used to date fossils. The nuclides we use in nuclear medicine, however, are not naturally occurring but rather are made either by bombarding stable atoms or by splitting massive atoms. There are three basic types of equipment that are used to make medical nuclides: generators, cyclotrons, and nuclear reactors.

Generators

Generators are units that contain a radioactive **"parent" nuclide** with a relatively long half-life that decays to a short-lived **"daughter" nuclide**. The most commonly used generator is the technetium-99m (^{99m}Tc) generator (Fig. 3-1), which consists of a heavily shielded column with molybdenum-99 (^{99}Mo; parent) bound to the alumina of the column. The ^{99m}Tc (daughter) is "milked" (eluted) by drawing sterile saline through the column into the vacuum vial. The parent ^{99}Mo (small grey circles) remains on the column, but the daughter ^{99m}Tc (white circles) is washed away in the saline.

A generator like the one just described is frequently called a **cow**, the elution of the daughter nuclide is referred to as **milking**, and the surrounding lead is called a **pig**, a term used for any crude cast-metal container. Generators come in

small sizes for use in a standard nuclear medicine department or in larger sizes for use in central laboratories.

Table 3-1 describes the features of three common generators.

Activity Curves for Generators

The plot of the curve describing the amount of daughter nuclide in a generator has two segments. The first traces the period of rapid accumulation of the daughter nuclide following creation of the generator or following **elution** (removal) of a portion of the daughter nuclide. This part of the curve lasts for approximately four half-lives of the daughter nuclide (which

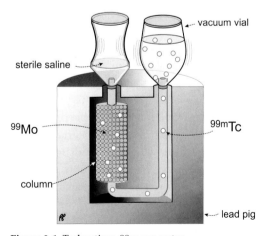

Figure 3-1 Technetium-99m generator.

Table 3-1 Characteristics of Three Commonly Used Generators

Generator (Parent–Daughter)	Clinical Uses of Daughter Nuclide	Half-Life of Parent ($T_{1/2p}$)	Half-Life of Daughter ($T_{1/2d}$)	$T_{1/2p}/$ $T_{1/2d}$
^{99}Mo–^{99m}Tc (molybdenum-99–technetium-99m)	Used in most radiopharmaceuticals for nuclear medicine studies	67 h	6 h	11.2
^{82}Sr–^{82}Rb (strontium-82–rubidium-82)	Cardiac perfusion imaging	600 h	0.021 h (75 s)	28 571
^{81}Rb–^{81m}Kr (rubidium-81–krypton-81m)	Lung ventilation scans	4.7 h	0.004 h (13 s)	1175

for ^{99m}Tc is approximately 24 hours). The second segment of the curve traces what is called the period of **equilibrium**, during which time the amount of daughter nuclide decreases as the parent nuclide decays.

During equilibrium, the activity of the daughter nuclide is described by the general equation [1]

$$A_d(t) = \frac{\lambda_d}{\lambda_d - \lambda_p} \times A_p(0) \times (e^{-\lambda_p t} - e^{-\lambda_d t})$$

$$(Eq.\ 3\text{-}1)$$

where t = time, $A_d(t)$ = Activity of daughter at time t, λ_d = decay constant for the daughter, λ_p = decay constant for the parent, $A_p(0)$ = activity of the parent at time 0. Substituting $0.693/T_{1/2}$ for λ yields

$$A_d(t) = \frac{0.693/T_{1/2d}}{(0.693/T_{1/2d}) - (0.693/T_{1/2p})} \times A_p(0)$$
$$\times (e^{(-0.693/T_{1/2p})t} - e^{-(0.693/T_{1/2d})t})$$

$$(Eq.\ 3\text{-}2)$$

Fortunately this calculation can be simplified by classifying generators into two groups: those in which the parent half-life is 10 to 100 times that of the daughter and those in which the parent half-life is more than 100 times that of the daughter. In the first group, the activity of the daughter during equilibrium decreases perceptibly over time. This is called **transient equilibrium**. On the other hand, the equilibrium segment of the curve for the second group is relatively flat. This is called **secular equilibrium**.

Transient Equilibrium

After a long period of time $e^{-\lambda_p t}$ is much greater than $e^{-\lambda_d t}$, and the latter value can be removed from Equation 3-1.

$$A_d(t) = \frac{\lambda_d}{(\lambda_d - \lambda_p)} \times A_p(0) \times (e^{-\lambda_p t}) \quad (Eq.\ 3\text{-}3)$$

By definition (see Eq. 1-4)

$$A_p(t) = A_p(0) \times (e^{-\lambda_p t})$$

Therefore, the equation describing the amount of activity of the daughter nuclide during transient equilibrium becomes

$$A_d(t) = \frac{\lambda_d}{\lambda_d - \lambda_p} \times A_p(t) \quad (Eq.\ 3\text{-}4)$$

or

$$A_d(t) = \frac{0.693/T_{1/2d}}{(0.693/T_{1/2d}) - (0.693/T_{1/2p})} \times A_p(t)$$

$$(Eq.\ 3\text{-}5)$$

Since $T_{1/2p}$ is greater than $T_{1/2d}$, then λ_d is greater than λ_p, and the ratio $\lambda_d/(\lambda_d - \lambda_p)$ will always be greater than 1; the activity of the daughter will be greater than the activity of the parent during equilibrium if 100% of the parent nuclide decays to the daughter nuclide. Transient equilibrium is illustrated in Figure 3-2. In this example, the half-life of the parent nuclide is approximately ten times that of the daughter. Following an elution that removes all of the available daughter, the amount of the daughter nuclide

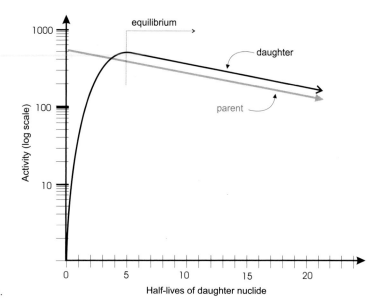

Figure 3-2 Transient equilibrium.

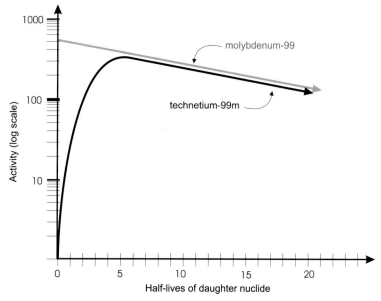

Figure 3-3 Transient equilibrium in a ^{99}Mo–^{99m}Tc generator.

rapidly increases until the daughter activity slightly exceeds that of the parent at about four to five half-lives. Thereafter the daughter activity declines at the same rate as the parent.

Only 86% of the molybdenum-99 decays to technetium-99m; the remainder decays directly to technetium-99 (Fig. 3-3). For the ^{99}Mo–^{99m}Tc

generator, Equation 3-4 becomes

$$A_{^{99m}\text{Tc}}(t) = 0.86 \times \frac{\lambda_{^{99m}\text{Tc}}}{(\lambda_{^{99m}\text{Tc}} - \lambda_{^{99m}\text{Mo}})} \times A_{^{99}\text{Mo}}(t)$$

$$= 0.86 \times \frac{0.116}{(0.116 - 0.010)} \times A_{^{99}\text{Mo}}(t)$$

$$= 0.94 \times A_{^{99}\text{Mo}}(t) \qquad (Eq.\ 3\text{-}6)$$

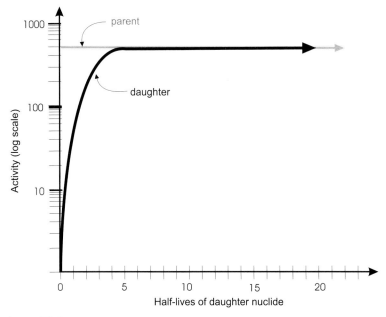

Figure 3-4 Secular equilibrium.

The activity of ^{99m}Tc is always less than the activity of ^{99}Mo (see Fig. 3-3).

Secular Equilibrium

For generators where the half-life of the parent is greater than 100 times that of the daughter nuclide, not only is $e^{-\lambda_p t}$ much greater than $e^{-\lambda_d t}$ (after a time period equivalent to several half-lives of the daughter nuclide) but λ_d is much greater than λ_p. Consequently, both λ_p and $e^{-\lambda_d t}$ can be removed from Equation 3-1 and at equilibrium the activity of the daughter nuclide equals the activity of the parent nuclide (Fig. 3-4).

$$A_d(t) = A_p(t) \qquad (Eq. 3\text{-}7)$$

Secular equilibrium, like transient equilibrium, is achieved rapidly following an elution that removes all of the available daughter nuclide. Thereafter the activity of the daughter nuclide is approximately equal to that of the parent. However, the decay curve of the parent appears to be flat since its half-life is so much longer than that of the daughter nuclide.

KINETIC ENERGY

Kinetic means "motion." The form of energy attributable to the motion of an object is its kinetic energy. Kinetic energy is related to both the mass (m) and velocity (v) of the object, specifically $1/2\ mv^2$. A moving car has kinetic energy, a parked car does not. A speeding car contains a great deal of kinetic energy that can be dissipated rapidly as heat, noise, and the destruction of metal in a collision.

Cyclotrons

Cyclotrons are circular devices (Fig. 3-5) in which charged particles such as protons and alpha particles are accelerated in a spiral path within a vacuum. The power supply provides a rapidly alternating voltage across the **dees** (the two halves of the circle). This produces a rapidly alternating electric field between the dees that accelerates the particles, which quickly acquire high kinetic energies. They spiral outward under the influence of the magnetic field until they have sufficient velocity and are deflected into a target.

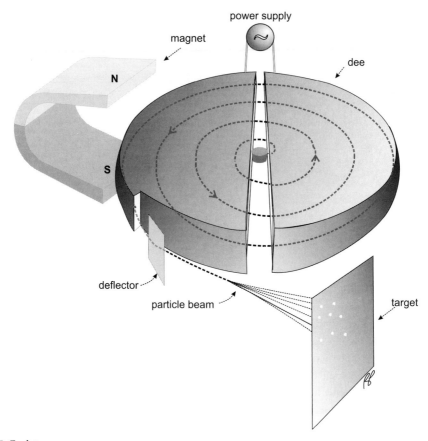

Figure 3-5 Cyclotron.

A deflector is used to direct the particles out through a window of the cyclotron into a **target**. Some of the particles and kinetic energy from these particles are incorporated into the nuclei of the atoms of the target. These energized (excited) nuclei are unstable.

Indium-111 (^{111}In) is produced in a cyclotron. The accelerated (bombarding) particles are protons. The target atoms are cadmium-111 (^{111}Cd). When a proton enters the nucleus of a ^{111}Cd atom, the ^{111}Cd is transformed into ^{111}In by discharging a neutron. This reaction can be written as:

Target atom (bombarding particle, emitted particle) product nuclide,

Cadmium-111(proton, neutron)Indium-111,

or

$$^{111}Cd(p,n)^{111}In.$$

Other examples of cyclotron reactions include $^{121}Sb(\alpha,2n)^{123}I$, $^{68}Zn(p,2n)^{67}Ga$, and $^{10}B(d,n)^{11}C$.

Reactors

Radionuclides for nuclear medicine are also produced in nuclear reactors. Some examples include ^{131}I, ^{133}Xe, and ^{99}Mo.

Reactor Basics

A general schematic of a reactor is depicted in Figure 3-6. It is composed of **fuel rods** that contain large atoms (typically uranium-235, uranium-238, or plutonium-239) that are

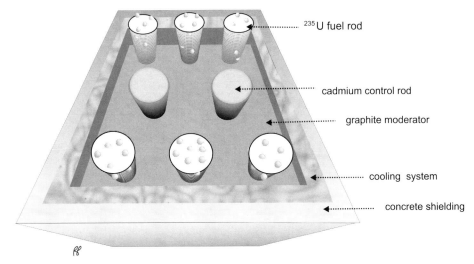

235U fuel rod

cadmium control rod

graphite moderator

cooling system

concrete shielding

Figure 3-6 Schematic of a nuclear reactor.

inherently unstable. These atoms undergo **fission** (see Fig. 1-16). Two or three neutrons and approximately 200 MeV of heat energy are emitted during this process. These neutrons leave the nucleus with moderately high kinetic energy and are referred to as **fast neutrons**. The neutrons are slowed with a **moderator** such as graphite, water, or heavy water. These "very slow" or **thermal neutrons**, and to a lesser extent the fast neutrons, in turn impact other fissionable atoms causing their fission, and so forth (Fig. 3-7). If this **chain reaction** were to grow unchecked, the mass would explode. To maintain control, cadmium **control rods** are inserted to absorb the neutrons in the reactor. They can be further inserted or withdrawn to control the speed of the reaction.

Medical nuclides are made in reactors by the processes of fission or neutron capture.

Fission

In this process, the desired radionuclide is one of the **fission fragments** of a heavy element ($Z > 92$), either the fuel atom itself or the atoms of a **target** placed inside the reactor. The **by-product** is chemically separated from the other fission fragments. The fission reaction is denoted as

^{235}Uranium (neutron, fission) daughter nuclide

For example, the formation of Iodine-131 and Molybdenum-99 are written as

^{235}U(n,f)^{131}I and ^{235}U(n,f)^{99}Mo

Neutron Capture

In **neutron capture** the target atom captures a neutron. The new atom is radioactive and emits gamma photons or charged particles to produce the daughter nuclide (Fig. 3-8). A gamma photon is emitted following capture of a thermal neutron. This reaction is written as

Target (neutron, gamma) daughter nuclide

For example,

^{98}Mo(n, γ) ^{99}Mo

When the target atom captures a fast neutron a proton can be emitted. This capture reaction is sometimes referred to as **transmutation** and is symbolized as

Target (neutron, proton) daughter nuclide

For example,

^{32}S(n,p)^{32}P

A list of common medical nuclides and their methods of production, modes of decay, and decay products is provided in Appendix A.

spontaneous fission of ^{235}U atom

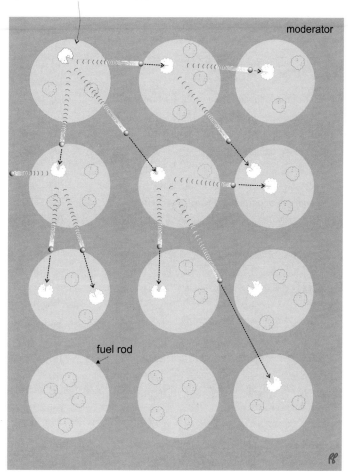

Figure 3-7 Chain reaction involving ^{235}U and slow neutrons.

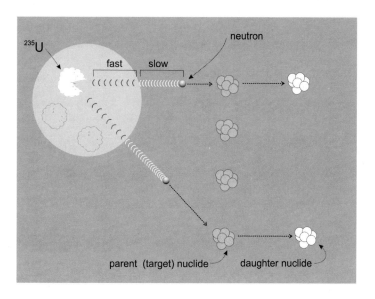

Figure 3-8 Neutron capture.

Reference

1 Saha, GB, *Fundamentals of Nuclear Pharmacy*, 5th Edition, Springer, New York, 2004, p. 22.

Test Yourself

1 Which of the following statements are true about radionuclide generators?

(a) The parent nuclide always has a shorter half-life than the daughter nuclide.

(b) If the $T_{1/2}$ of the parent nuclide is 50 times greater than the $T_{1/2}$ of the daughter nuclide the equilibrium portion of the activity curve is basically flat and is categorized as "secular" equilibrium.

(c) The parent nuclide is less tightly bound to the column than the daughter nuclide.

(d) All of the above.

(e) None of the above.

2 True or false: During an equilibrium state within a ^{99}Mo–^{99m}Tc generator the total activity of ^{99m}Tc is always less than the total activity of ^{99}Mo because 14% of ^{99}Mo decays directly to ^{99}Tc, bypassing the metastable state.

3 Associate each of the following terms (a) through (k) with the most appropriate of the three methods of nuclide production listed here (a term can be applied to more than one method):

(a) Moderator

(b) Chain reaction

(c) Thermal neutron

(d) Dee

(e) Control rod

(f) Target

(g) Cow

(h) Pig

(i) Column

(j) Elution

(k) By-product

  (i) Generator

  (ii) Cyclotron

  (iii) Reactor.

4 Associate each of the following nuclear reactions with the most appropriate of the three, listed production methods:

(a) ^{111}Cd(p,n)^{111}In

(b) ^{68}Zn(p,2n)^{67}Ga

(c) ^{235}U(n,f)^{99}Mo

(d) ^{98}Mo(n,γ)^{99}Mo

  (i) Cyclotron

  (ii) Fission reaction (reactor)

  (iii) Neutron capture (reactor).

4 Nonscintillation detectors

Electronic equipment has been developed to detect ionizing radiation (both particles and photons) as we generally cannot sense the presence of radioactivity. This chapter explores the common types of nonscintillation radiation detectors used in a nuclear medicine department; Chapter 5 discusses scintillation detectors.

Gas-Filled Detectors

Theory of Operation

Gas-filled detectors function by measuring the ionization that radiation produces within the gas. There are several types of detectors that operate on this general principle, but they differ greatly in the details of construction and in the manner in which the radiation-produced ionization is gauged. As might be expected, each type has one or more applications for which it is best suited.

One important factor determining the applicability of gas-filled detectors is the nature and state of the gas itself. For example, because the molecules of any gas are relatively widely separated, they are more likely to be ionized by the strongly ionizing charged particles, such as alpha and beta radiation, than by gamma or photon radiation. Partially for this reason, a gas-filled detector is usually used to monitor alpha and beta radiation, although with

appropriate design a gas-filled detector can also be used to measure photon (gamma or x-ray) radiation.

Principles of Measurement

Charge Neutralization

Perhaps one of the simplest and oldest methods for quantifying the ionization produced in a gas detector is the visual observation of the filament of an electroscope. An example is the pocket dosimeter, illustrated later in this chapter in Figure 4-10, which contains a small gas-filled chamber within which a thin filament of wire is attached to a metal frame. When positively charged, the filament is repelled by the frame, which also carries a positive charge. The filament is observed to return to its resting position as the charge is neutralized by the radiation-produced ions. The greater the incoming radiation, the closer the filament moves toward the neutral position.

Charge Flow

Measuring Current

A related approach is to measure the flow of the charges that ionizing radiation produces in a gas-filled detector. The ions produced by the radiation are charged particles. The negative particle is either a free electron or an oxygen or

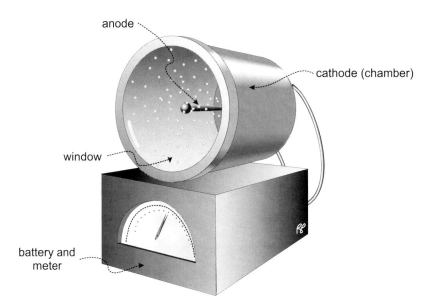

Figure 4-1 Simple gas-filled detector.

nitrogen molecule that has absorbed a free electron. The positive particle is a molecule of gas that has lost an electron.

The gas-filled detector has both a positive and a negative electrode (Fig. 4-1). As shown in a cross-section in Figure 4-2, the potential difference between them is maintained by a battery. The positive and negative ions produced in the gas by the radiation move in opposite directions, positive ions toward the negative cathode and the negative ions toward the anode. This movement of ions (charges) is an electric current, which can be detected by a sensitive meter. The current between the electrodes is a measure of the amount of incoming radiation. The **ionization chamber**, about which more is said later in this chapter, is a practical instrument that functions in this way.

Counting Pulses of Current

The alternative to measuring the current is counting the individual pulses produced as each individual charged particle or photon enters the gas. The **Geiger counter** is an example of this type of detector. The rate at which counts occur in such detectors is a direct measure of the amount of incoming radiation.

The process of ionization, collection of the charges produced, and recording of the count takes place very quickly, but it is far from instantaneous. Time is required for the ions to reach their respective electrodes and for the detector to return to its resting state. What transpires in this time depends on the construction of the detector, the kind of gases it contains, and the strength of the applied voltage. The last of these is almost directly proportional to the voltage applied to the electrodes. Although all three factors determine the characteristics of the detector, it is instructive to examine the role of the applied voltage in more detail.

Characteristics of the Major Voltage Regions Applied Across a Gas-Filled Detector

Low

In the gas-filled detector, the magnitude of the voltage between the electrodes determines the type of response to each charged particle or photon. When the voltage between the electrodes is relatively low, the field within the gas is weak and many of the ions simply recombine, leaving only a small fraction to reach the electrodes. Little if

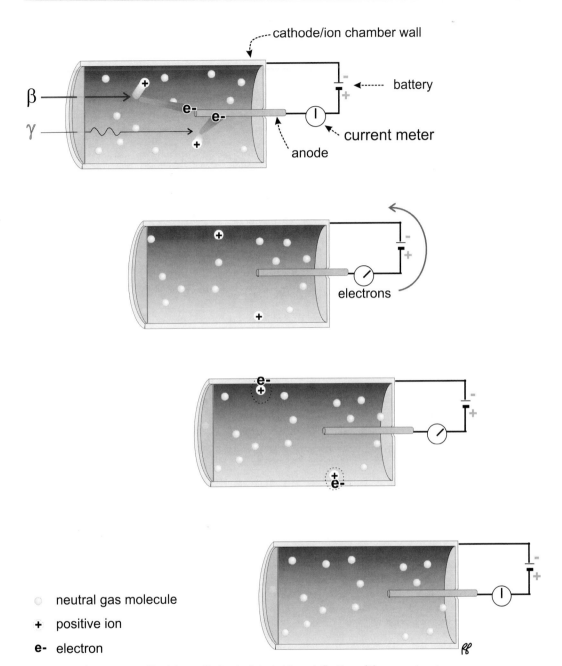

Figure 4-2 The presence of ionizing radiation is detected by a deflection of the current meter.

any charge flows between the electrodes and the meter in the external circuit remains at zero.

At a somewhat higher voltage, referred to as the **ionization region** (Fig. 4-3), most of the ions that are formed reach the electrodes. A further small increase in the voltage does not increase the current once the voltage is sufficient to collect 100% of the ions formed. The brief pulse of current, generated by ionizing radiation entering the chamber, ceases until the next charged particle or photon enters the gas. This current is small and difficult to count as an individual event.

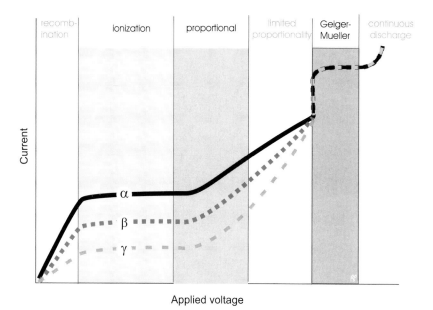

Figure 4-3 Current as a function of applied voltage in a gas detector. The regions of interest include ionization, proportional, and Geiger.

In this region, the current is relatively independent of small increases in voltage. It is, however, affected by the type of radiation. An alpha particle, because it carries two units of charge and is relatively massive, produces many ion pairs while traveling a short distance in the gas; a beta particle, which is much lighter and carries only a single charge, produces fewer ion pairs per unit distance traveled; a photon, because it carries neither charge nor mass, creates even fewer ion pairs. In any case, practically all of the ion pairs that are created are collected on the electrodes because the applied voltage creates a strong enough electric field to prevent recombination.

In small detectors, some of the ionization from beta radiation and much or even most of the ionization from photon radiation may escape detection. In this situation, the current following irradiation by an alpha particle will be much larger than that caused by a beta particle of similar energy, and the current following both particles will be for both larger than that caused by photon radiation of similar energy.

Intermediate

With further increase in voltage, the detector passes into the next region of operation (see Fig. 4-3). In this so-called **proportional region**, a new phenomenon is observed—**gas amplification** (Fig. 4-4). Accelerated more intensely toward the positive electrode at this higher voltage, the electrons produced by the radiation (called primary particles) travel so rapidly that they themselves are able to ionize some of the previously neutral gas molecules. This process, similar to that produced by the original radiation particle, can be imagined as a speeding electron knocking out a molecular electron (the new negative ion) and leaving behind a positively charged gas molecule (the new positive ion). The newly separated electrons (called **delta** or secondary particles) also accelerate toward the positive electrode and, in turn, ionize additional gas molecules. The pulse of charge started by the incoming radiation is greatly amplified by this brief chain reaction.

In this region, the current produced is proportional to the number of ion pairs produced by

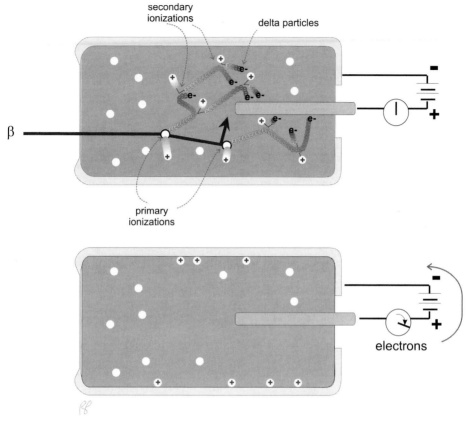

Figure 4-4 Proportional counter. The voltage causes gas amplification that gives electrons separated during primary ionization enough energy to cause secondary ionization.

the incoming radiation. The current is higher for an alpha particle than for a beta, and currents for both are higher than for a photon. Because of gas amplification, the total number of ion pairs, primary plus secondary, is much higher than in the low-voltage ionization region. This resulting pulse of current is large enough to detect as an individual, countable event.

High

For detectors operating at still higher voltages, above the proportional region, the pulse of current is larger but becomes independent of the number of ions produced by the initial event. As the voltage is increased, a point is reached at which most of the gas within the detector is massively involved in the multiple, successive ionizations (Fig. 4-5). Once all the gas is

involved, no greater gas amplification is possible so that any further increases in voltage have little effect on the size of the pulse of current. This is the so-called **Geiger region**. A detector operating in this region is called a Geiger counter, or Geiger-Mueller counter, after its early developers. In the Geiger region, not only is the size of the current pulse almost independent of small changes in voltage, but the size of the pulse is also independent of the amount of ionization produced by the incoming radiation. In other words, in the Geiger region the current produced by a charged particle or photon is large compared to that produced in the proportional region. The current is independent of fluctuations in voltage, and the size of each pulse of current is dependent on the characteristic of the detector itself rather than of the incoming particle or photon.

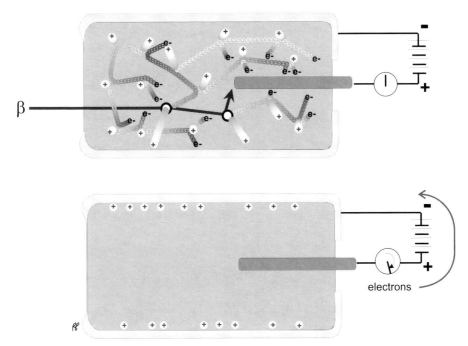

Figure 4-5 Geiger counter. The primary radiation rapidly triggers a cascade of further ionizations involving most of the gas.

Figure 4-6 Continuous discharge. The voltage applied to a neon sign is high enough to generate spontaneous ionization that continues until the power is turned off.

Voltages above the Geiger region are not used because, even in the absence of the radiation the counter is designed to detect, there is a spontaneous and **continuous ionization** of gas molecules that stops only when the voltage is lowered. This is similar to the visible ionizations seen in a neon sign (Fig. 4-6). With such high voltages, the device is not useful as a radiation detector.

Sensitivity

Intrinsic

A gas-filled detector will respond to virtually every radiation event that causes ionization in

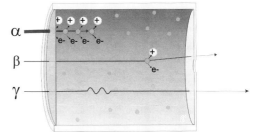

Figure 4-7 Ionization by alpha and beta particles and by photons in the gas-filled detector.

the gas. To be detected, a particle or photon must be energetic enough to cross the detector's face into the sensitive volume of gas, but must not be so energetic that it will pass right through the gas without causing any ionization. The first limitation is important for low-energy alpha or beta particles, which have only limited ability to penetrate the "window" of the detector but once inside ionize strongly. The second is a consideration for high-energy photons, which penetrate easily but may pass through the detector causing little or no ionization (Fig. 4-7).

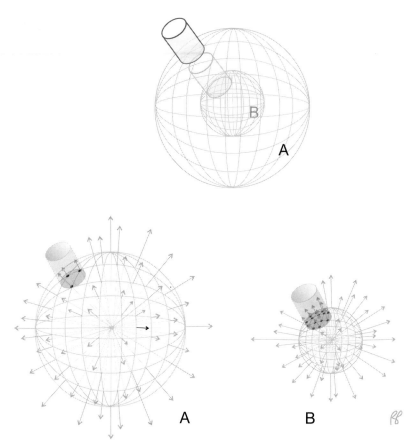

Figure 4-8 Geometric efficiency of detectors. The closer the detector is to the source, the greater the number of photons that will cross the detector window.

Sensitivity for charged particles (alpha or beta) can be increased by using thin, penetrable materials for the detector window such as a thin sheet of mica or, for greatest sensitivity, by actually placing the radioactive sample inside the sensitive volume. Sensitivity for moderately high-energy photons is improved by increasing the size of the sensitive volume or, equivalently, by "cramming" in more gas molecules under pressure or by using a thicker window.

Typical gas-filled detectors are sensitive to alpha particles with energies greater than 3 MeV to 4 MeV, beta radiation above 50 keV to 100 keV, and gamma radiation above 5 keV to 7 keV. The upper limit for gamma or x-ray radiation depends on the type and pressure of the gas and the type of material used for the window and walls of the detector.

Geometric

The larger the window and the closer the source, the more radiation will enter the detector. At long distances from compact sources, the amount of radiation reaching the detector decreases with the square of the distance. The controlling factor, aside from absorption of radiation in the intervening air, is the portion of the source seen by the window. As can be seen in Figure 4-8, this is the fraction of the sphere subtended (or seen) by the window.

Types of Gas-Filled Detectors

Ionization Chambers

Structure and Characteristics

Structure: The ionization chamber, in its simplest form, is a gas-filled can (the gas is usually air) with a radiation permeable end (the window), a central wire, a meter, and a battery (see Fig. 4-1). The gas is most often at normal atmospheric pressure, but it may be filled under pressure, as discussed in the section on Geiger counters. A thin window at one end of the can admits radiation. A battery or power supply maintains the central wire at a positive potential relative to the surrounding walls of the can, which act as the negative electrode. The magnitude of the voltage is set relatively low to ensure operation in the ionization region, hence the name **ionization chamber** (see Fig. 4-3). A sensitive meter to measure the current of ions completes the device.

Function: The ionization chamber is not ordinarily used to count discrete radiation events but rather to measure the average number of ionizations per minute occurring within the gas. If the events are sufficiently infrequent, the attached meter may be observed to move up a short distance as a particle or photon of radiation triggers the detector and falls back slowly in the interval before the event.

Sensitivity: The lower limit of sensitivity for an ionization chamber is determined by the sensitivity of the meter used to measure the current. In terms of radiation exposure, sensitivity down to <1 mR for low- and moderate-energy photons (10 keV to 1 MeV) is available in standard survey meters and dosimeters.

Energy independence: In the ionization region of operation the electrodes collect practically all of the ion pairs formed in the gas. Because the number of ionizations is almost directly proportional to the energy of the incoming radiation, the current that results is a measure of the rate at which energy is being deposited within the gas by the ionizing radiation. Readings are expressed in rads or sieverts per hour. The rate of ionization, the consequent current, and the radiation dose are all similarly dependent on the energy of the incoming particles or photons. As a result, at least for the ideal ionization chamber, the meter reading gives a reliable measure of the radiation dose rate. Note, however, that this reliability has its limits. As alluded to above, at very low energies the radiation may not even penetrate the ion chamber, and at very high energies the radiation may pass completely through the device without having many ionizing interactions with the gas.

Applications

Dose calibrator: The dose calibrator is most frequently used in the nuclear medicine department as a table-top ionization chamber to confirm that the correct amount of activity has been dispensed before a dose of radiopharmaceutical is administered (Fig. 4-9). The dose calibrator consists of an ionization chamber surrounding an open well. The walls of the well are permeable to photons. The current produced in the circuitry is proportional to the number of primary ionizations in the chamber. The amount of current is registered as radioactivity in megabecquerel or millicurie. The dose calibrator can only report the activity, not the type of radiation or radiopharmaceutical.

The accuracy of the reading is affected by such factors as the type of dose container, its proper placement in the dose calibrator, and the calibration and regular recalibration of the instrument itself. Because of the importance of administering the correct amount of activity to patients, there are strictly enforced rules for the use and calibration of the dose calibrator. These are discussed in Chapter 10.

Survey meter: When the ionization chamber is used as a survey meter, the current reading is usually interpreted as the average intensity of radiation in roentgens (R) per hour. For example,

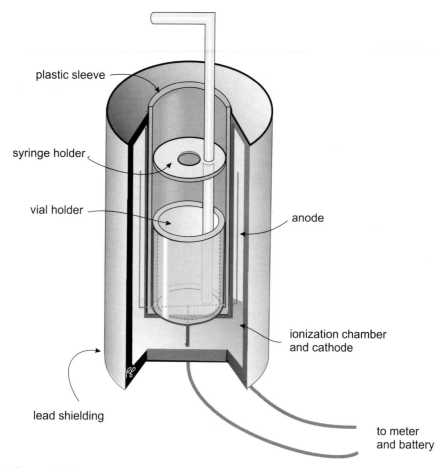

plastic sleeve

syringe holder

vial holder

anode

ionization chamber
and cathode

lead shielding

to meter
and battery

Figure 4-9 Dose calibrator.

a survey meter might register a 30 mR/hr exposure rate at 1 m from a person who was treated with 370 MBq (10 mCi) of ^{131}I. The roentgen is defined in terms of ionization produced in air and it is no coincidence that the ionization chamber can be used for the measurement of radiation intensity in roentgens.

ROENTGEN (R)

A measurement of gamma or x-ray radiation exposure in air, not tissue. One roentgen is the quantity of radiation that will produce 2.58×10^{-4} coulombs (C) of charge (or 2×10^9 ion pairs) per kilogram of dry air under standard conditions.

Pocket dosimeters: A small ionization chamber is the heart of the classic **pocket dosimeter**. For this application, a small, straight filament, insulated from the walls of the chamber, is mounted within the ionization chamber (Fig. 4-10). To prepare the chamber for use, a positive charge is placed on the filament by a charger that briefly connects the positive terminal of a battery to the filament. When charged, the positive filament is repelled by the positively charged frame. When radiation penetrates the walls of the chamber, the gas is ionized and the ions are attracted to the frame and fiber or walls of the chamber, which partially neutralizes their charge. The filament is less strongly repelled by the frame and begins to move toward a neutral position. The position of the filament

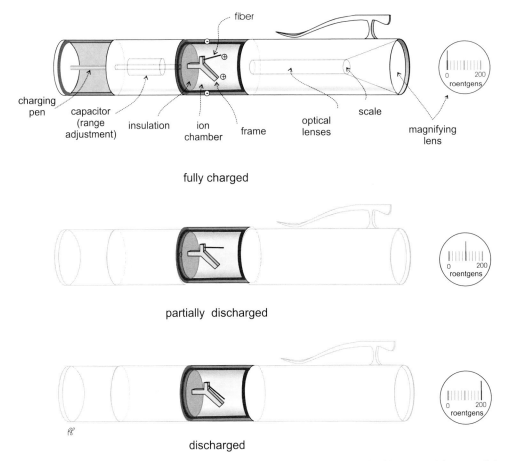

fiber

charging
pen

capacitor
(range
adjustment)

insulation

ion
chamber

frame

optical
lenses

scale

magnifying
lens

0   200
roentgens

fully charged

0   200
roentgens

partially discharged

0   200
roentgens

discharged

Figure 4-10 Pocket dosimeter. Ions produced in the gas neutralize the charge on the filament and frame and the filament returns toward its neutral position. (Adapted from drawing of the FEMA dosimeter, courtesy of FEMA.)

can be viewed through a lens against a scale on the end of the chamber calibrated in roentgens or rads per hour. Properly insulated from the case of the chamber, the positive charge can remain on the filament for hours.

Pocket dosimeters of this type are relatively inexpensive and can be used to measure exposures to photons in the range of zero to several hundred milliroentgen. A separate charging device is required.

Proportional Counters

Structure and Characteristics

Chamber and filling gas: In its construction, the proportional counter is very similar to an ionization chamber. The filling gas is more likely to be argon or an argon–methane mixture than air, but the essential difference between them is the higher voltage and the resulting gas amplification. For this reason the incoming radiation causes a pulse of current large enough to be registered as an individual count.

Applications as Survey Meter

The **proportional counter** is particularly suitable when it is important to distinguish among various types of radiation. Because the size of the pulse from a proportional counter is proportional to the initial ionization, alpha particles, which ionize more heavily than beta particles, produce

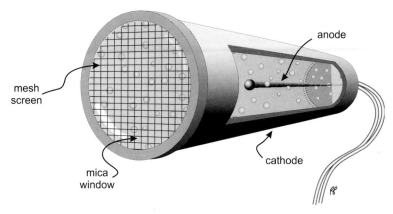

Figure 4-11 Geiger probe.

larger pulses, and this difference can be used to distinguish between them.

· Counting low-energy particles may require very close proximity of the detector to the sample. Proportional counters can be built for this type of counting with no window or only a very thin, somewhat leaky membrane between the sample and the gas of the chamber. In the windowless type, the sample is placed directly within the gas-filled chamber. A continuous flow of gas through the detector compensates for any loss of gas as the samples are inserted or, in the case of the thin window, for any gas that leaks through the membrane. Gas-flow counting is relatively exacting and, for many purposes, has been replaced by the less exacting liquid scintillation counting.

Geiger Counters

Structure and Characteristics

The tube and the filling gas: The **Geiger counter** is an ionization chamber that operates at a relatively high applied voltage (Fig. 4-11). The chamber is usually filled with argon containing traces of other gases such as halogen or methane, although a detector will function with a filling as simple as dry air. The sensitive end of the probe is generally a mica window protected by an external metal mesh. In some applications, the thin window of the ionization chamber is covered by an aluminum cap. Photons striking the cap knock

out secondary electrons that in turn ionize the gas within the chamber. The walls of the chamber may also function similarly.

The gas in the chamber may be filled to atmospheric pressure or it may be pressurized to increase sensitivity. At higher pressure, usually a few times the normal atmospheric pressure, the number of gas molecules "crammed" into the chamber is greater. This raises the probability that incoming radiation will encounter and ionize a gas molecule within the chamber. Pressurization is particularly useful for photon radiation for which the energy is high enough, and therefore its range in gas long enough, to allow it to pass completely through an unpressurized chamber without encountering and ionizing a gas molecule. The Geiger counter cannot distinguish between types of radiation because each interaction of the radiation with the gas causes maximum ionization.

Quenching

The gas multiplication, which characterizes the Geiger region, causes a single incoming radiation event to give a large current pulse, but gas amplification carries the disadvantage that the discharge, once started by the radiation, is likely to sustain itself indefinitely. While the discharge continues, the detector is not affected by any further incoming radiation; it is effectively paralyzed (this **dead time** lasts 100–500 μs for Geiger

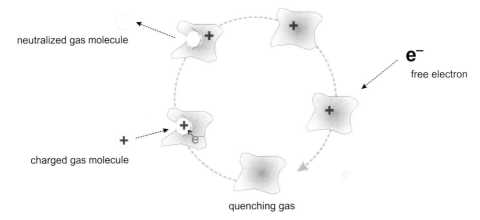

neutralized gas molecule

e⁻
free electron

charged gas molecule

quenching gas

Figure 4-12 Quenching gas used in Geiger probes.

counters). The discharge must be quenched before the tube can count again. The two common methods of **quenching** are to quickly drop the applied voltage or to add a quenching material to the filling gas. This material serves to quench the discharge by absorbing the kinetic energy of the electrons and facilitating their recombination with the positive ions (Fig. 4-12). Halogen compounds and small organic molecules such as methane gas are commonly used for this purpose. Whatever method is used, quenching is transparent to the user.

Applications

The Geiger counter has long been the most widely used of the gas-filled detectors. Its principal uses are the monitoring of areas such as nuclear medicine laboratories for radiation and the detection of contamination. When used as a **radiation monitor** to detect individual photons or their rate (usually referred to as counts or count rate, respectively), the Geiger counter is relatively independent of the energy of the photon. This is true provided that the photon is sufficiently energetic to enter the counting chamber but not so energetic that it passes through it without interacting. When used as a **survey meter** to measure the exposure rate, usually as milliroentgen per hour, the meter reading is strongly affected by the energy of the photons.

For low-energy photons, typically those below 100 keV, the actual exposure rate is only a fraction of the reading displayed on the meter. The reason for this is that the counter detects individual photons, not their energy, whereas the exposure rate depends both on the photon and the energy of the photon. The usual meter is calibrated for a photon of moderate or even high energy such as that of ^{137}Cs or ^{60}Co. It must be recalibrated if the energy of the photon is expected to be very different from that used for the factory calibration.

Sensitivity: The Geiger counter can be expected to respond to any individual particle or photon whose energy is high enough to permit it to penetrate into the chamber but low enough so that it does not, or is unlikely to, pass through without ionizing any gas molecules. From this, it is apparent that geometric considerations—the size and shape of the detector and its distance from the source—can be the most important factors limiting sensitivity.

For monitoring radiation of low intensity, sensitivity is further limited by the background count. The count rate of the monitored radiation must be sufficient to increase the total count rate significantly above background. For radiation of high intensity, sensitivity may be limited by the so-called dead time of the counter, which,

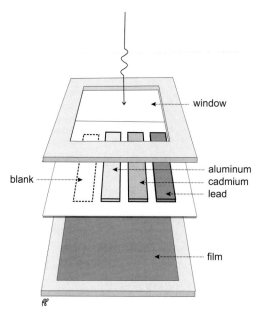

Figure 4-13 Film badge. The absorbers assist in estimating the type and energy of radiation exposure.

as described above, is the time before one discharge has been quenched and the tube is able to count another event.

Photographic Detectors

The **film badge** depicted in Figure 4-13 is a detector that is commonly used to measure the cumulative exposure received by personnel working with radioactivity. It is simply a plastic holder containing film that is radiosensitive. Strips of materials of different densities (such as aluminum, cadmium, and lead) are placed within the badge in the space in front of the film. These strips attenuate the incoming radiation and reduce the degree of exposure of the film located immediately behind the strip. As discussed in Chapter 2, the amount of attenuation is dependent on the type of radiation (alpha, beta, gamma,

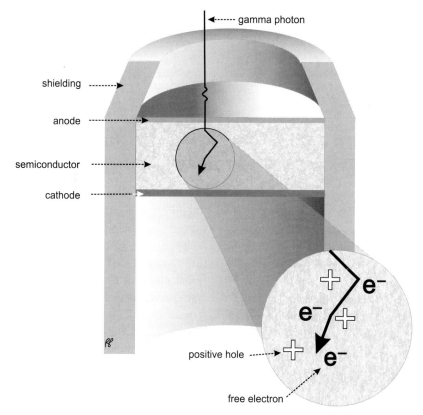

Figure 4-14 Semiconductor detector. Photons entering the detector create electron–positive hole pairs.

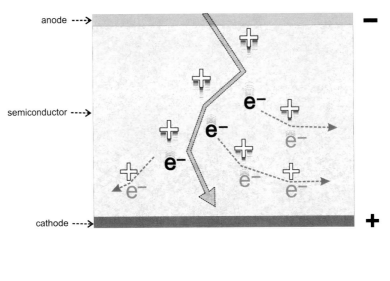

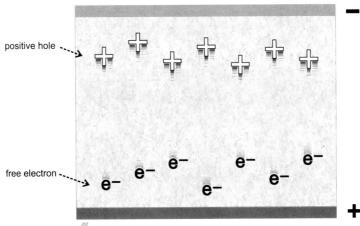

Figure 4-15 Electron and positive holes formed by the primary and secondary ionizations migrate in the field created by the anode and cathode.

or x-ray), the energy of the radiation, and the density of the absorber. By comparing the amount of exposure of film behind each strip (including a fourth uncovered area), some estimate as to the type and energy of the radiation can be determined. Unlike the pen dosimeter, which can be read by the user, the film badge must be sent to an outside laboratory for interpretation.

Semiconductor Detectors

Semiconductor detectors can be thought of as functionally equivalent to a gas detector but with all the advantages of a solid material. Like their gaseous counterparts, the electrons of the molecules of these solids can be "dislodged" by ionizing radiation. Instead of positively charged gas molecules, positive "holes" are created within the crystalline structure (Fig. 4-14). In a manner similar to their gaseous counterparts, for which the ion pairs of electron and positively charged gas molecules migrate within the gas chamber, the electrons and the positive "holes" migrate within the semiconductor. These charged entities are attracted to an anode and cathode attached to opposite sides of the detector (Fig. 4-15).

Semiconductor materials are less conductive than the metals for carrying electric current (such as copper), hence the name *semi*conductor.

The primary advantage of using solid materials instead of gas for detectors is their much higher density. The higher the density of a material the more likely the interaction of the incoming radiation with atoms of the material. In addition, the electrons in the semiconductor are less tightly bound to their atoms than the electrons in the atoms of gas molecules. It takes only 2 eV to 3 eV to "release" an electron in a semiconductor material compared to approximately 35 eV to release an electron in air. This means that for any incoming radiation there is a much greater yield of charges (positive holes and negative electrons) in the semiconductors than in air (or other gases) of a similar volume. As a consequence, a relatively smaller volume of solid material is needed, which allows the production of much smaller detectors.

A few of the many semiconductor compounds that are available for use include cadmium telluride (CdTe), cadmium zinc telluride (CdZnTe), and zinc telluride (ZnTe). Because of the high cost of manufacturing semiconductor materials, they are currently only used for smaller detector units (such as intraoperative probes). Once some of the manufacturing difficulties are resolved, it is expected that they will be available for use in imaging devices.

Test Yourself

1 True or false: Gas filled detectors are highly efficient for detection of high-energy gamma photons and x-rays as well as low-energy beta particles.

2 Which of the following radiation detectors are classified as ionization chambers?
   (a) Pen dosimeter
   (b) Geiger counter
   (c) Dose calibrator.

3 True or false: Because of the near-maximal ionization of the gas molecules in a Geiger counter in response to each radiation interacting with a gas molecule in the detector, Geiger counters cannot be used to distinguish between types or energies of radiation.

4 True or false: The film badge, unlike the pen dosimeter, can provide some information about the type and energy of the radiation an individual has received.

5 Nonimaging scintillation detectors

Structure and Characteristics of the Crystal Scintillation Detector

Principal Components

Crystal

A **crystal** is a clear slab in which gamma rays are converted to light. The most widely used crystals are made of sodium iodide (NaI); they are fragile and can easily be cracked. Because sodium iodide crystals absorb moisture from the atmosphere, they must be sealed in an air-tight aluminum container.

Sodium iodide crystals are "**doped**" with small amounts of stable thallium (Tl). The thallium atoms dispersed in the crystal improve its response to the gamma ray photons (Fig. 5-1). The process of converting gamma rays to light is complex, but it can be summarized as absorption of the gamma ray energy by the crystal, leaving its electrons in an excited state. The gamma photon transfers its energy in one or more Compton or photoelectric interactions in the crystal. Each of these energetic electrons produced by the gamma ray interactions in turn distributes its energy among electrons in the crystal leaving them in an excited state. As these return to their original state, some of their energy is released as **light photons** (Fig. 5-2).

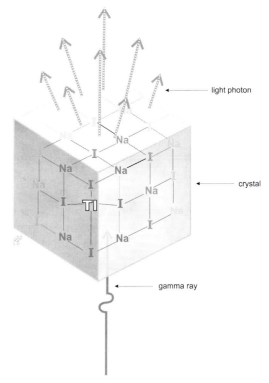

Figure 5-1 Scintillation crystal. The sodium iodide crystal "doped" with a thallium impurity is used to convert gamma photons into light photons.

For each kiloelectronvolt (keV) of gamma ray energy absorbed by the crystal, approximately forty light photons are emitted. Photomultiplier tubes detect these light photons.

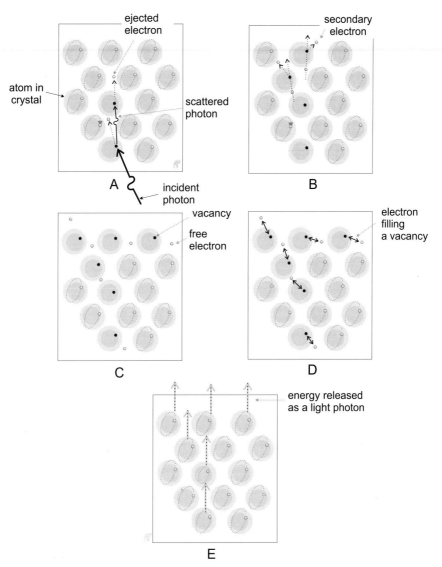

Figure 5-2 Light photons. (A) Gamma rays eject electrons from the crystal through Compton scattering and the photoelectric effect. (B, C) The ejected electrons in turn produce a large number of secondary electrons. (D, E) During de-excitation (oversimplified in this drawing) energy is released in visible light.

The design of the crystal affects its performance. The thickness of the sodium iodide crystal ranges from less than a centimeter to several centimeters. Thicker crystals, by absorbing more of the original and the scattered gamma rays, have a relatively high sensitivity in which almost all of the gamma ray energy reaching the crystal is absorbed (Fig. 5-3). Thinner crystals have lower sensitivity because more photons escape.

For photons in the 140 keV range (^{99m}Tc), typical thicknesses range from 0.6 cm to 1.2 cm.

Photomultiplier Tubes

The **photomultiplier tube** is a vacuum tube with a **photocathode** on the end, placed adjacent to the crystal. A photocathode is a clear photosensitive glass surface. This is coupled with a light-conductive transparent gel to the surface of the

Figure 5-3 Thick crystals stop a larger fraction of the photons.

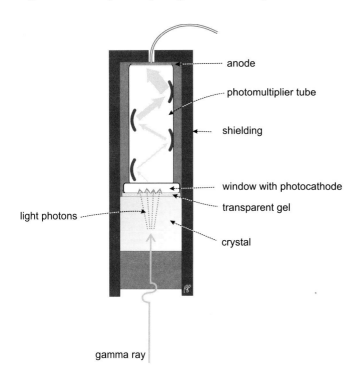

anode

photomultiplier tube

shielding

window with photocathode

transparent gel

crystal

light photons

gamma ray

Figure 5-4 Sodium iodide crystal scintillation detector.

crystal (Fig. 5-4). The transparent gel has the same refractive index as the crystal and the photomultiplier tube (PMT) window. The light striking the photocathode causes it to emit electrons, referred to as photoelectrons. On average, four to six light photons strike the photocathode for each photoelectron produced.

The number of electrons produced at the photocathode is greatly increased by the multiplying action within the tube (Fig. 5-5). As soon as they are produced, the electrons cascade along the multiplier portion of the tube successively striking each of the tube's **dynodes**. These are metal electrodes, each held at a progressively higher positive charge than the one before it. As an electron strikes a dynode, it knocks out two to four new electrons, each of which joins the progressively larger pulse of electrons cascading toward the anode at the end of the tube. In other words, for each electron entering a cascade of just three such dynodes, there will be between 2^3 and 4^3 electrons leaving; cascading against ten dynodes would yield between 2^{10} and 4^{10} electrons.

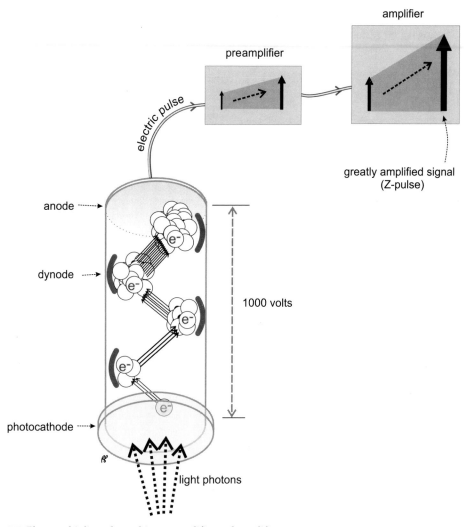

Figure 5-5 Photomultiplier tube and its preamplifier and amplifier.

Preamplifiers and Amplifiers

The current from the photomultiplier must be further amplified before it can be processed and counted. Despite the multiplication within the photomultiplier tube, the number of electrons yielded by the chain of events that begins with absorption of a single gamma ray in the crystal is still small and must be further increased or amplified. Typically, this amplification is a two-stage process.

In the first stage, a small **preamplifier** located close to the photomultiplier increases the number of charges sufficiently to allow the current to be transmitted through a cable to the main **amplifier**. In the second stage, the current of the electrical pulse is further increased by the main amplifier as much as a thousand-fold (see Fig. 5-5).

Pulse-Height Analyzer

The amplifiers are designed to ensure that the amplitude of each pulse is proportional to the energy absorbed in the crystal from the gamma radiation. The amplitude of each electrical pulse from the amplifiers is measured in the electrical circuits of the **pulse-height analyzer**. A tally is kept showing the number of pulses of each height. A plot of the number of pulses against their height—that is, their energy—is called a **pulse-height spectrum**.

The pulse-height analyzer is often used to "select" only pulses (conventionally called **Z-pulses**) that correspond to a range of the acceptable energies. This range is called the **energy window**. A window setting of 20% for the 140 keV photopeak of ^{99m}Tc means that Z-pulses corresponding to a 28 keV range centered on 140 keV (126 keV to 154 keV) will be accepted and counted (Fig. 5-6).

Sodium Iodide Detector Energy Spectrum

The shape of the pulse-height spectrum is dependent on the photon energies and the characteristics of the crystal detector. We shall now review

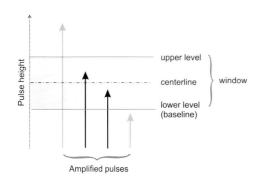

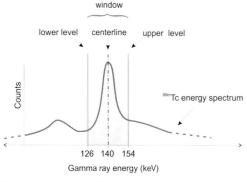

Figure 5-6 Pulse-height analyzer. The incoming pulse (Z-pulse) is proportional to the energy of the initial gamma ray photon. The pulse-height analyzer accepts only those that fall within the window.

some of the general features of the spectrum and examine the components of the spectrum of a NaI crystal, the most commonly used crystal.

CALIBRATING THE ENERGY SPECTRUM

The energy scale (horizontal axis) can be calibrated in absolute terms by using a radionuclide whose photon energy is well known; usually one with a simple decay scheme. The most prominent peak, seen as a high point in the spectrum, is then assigned the energy value corresponding to the known energy of the principal gamma ray of the radionuclide under observation.

Photopeak

When a gamma photon deposits all of its energy in the crystal, the amplifier output is a single electrical pulse whose amplitude is, as discussed above, proportional to the energy of the original gamma photon. Ideally this conversion of gamma energy to electrical pulse would be identical for each photon, and a plot of these pulses would appear as a single narrow "spike." However, for a variety of reasons, such as the physical variations and the minor imperfections in the process of collecting and converting light photons into electric current, the plot of electrical pulses corresponding to the photon energy is only a statistically blurred version of the original spike (Fig. 5-7).

Photopeaks in the spectrum correspond to the principal energies of gamma rays from the radioactive source. Figure 5-8 shows a typical spectrum, one from a sample of ^{99m}Tc. The relatively sharp, prominent peak at 140 keV is called the photopeak. It is produced by the prominent gamma photon of ^{99m}Tc. The horizontal axis of the pulse-height spectrum represents energy, typically in kiloelectronvolt (keV) or megaelectronvolt (MeV). The vertical axis represents the number of photons detected at each point on the energy scale.

Other Peaks in the Energy Spectrum of the Source

Several other important peaks—Compton, iodine escape, annihilation, and coincidence—are briefly discussed below.

Compton Peak (or Compton Edge)

Compton scattering was introduced in Chapter 2. If both the **Compton electron** and deflected

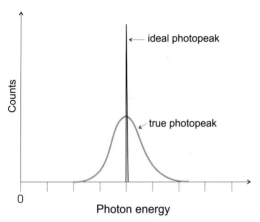

Figure 5-7 Imperfections in the crystal and circuitry cause blurring of the photopeak.

photon are detected their total energy will equal that of the incident photon, and the event will register in the photopeak. However, the scattered photon often escapes detection, so that the event leaves only the energy of the Compton electron in the crystal. These Compton electrons, whose energies are always less than that of the incident photon, register to the left of the photopeak.

The Compton electrons can have any energy from nearly zero up to a characteristic maximum called the **Compton peak** or **edge**. The value of the maximum energy can be calculated as follows:

$E_{\text{maximum compton electron}}$

$$= E^2_{\text{incident photon}}/(E_{\text{incident photon}} + 0.256)$$

(Eq. 5-1)

For example, the maximum Compton electron energy for a ^{99m}Tc 140 keV photon is 50 keV. The derivation of Equation 5-1 is outlined in the Appendix included as a final section to this chapter.

The **Compton plateau** refers to electron energies that are less than the Compton peak; the

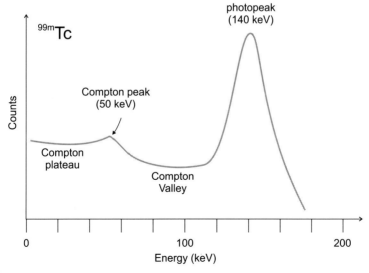

Figure 5-8 Compton peak. (Adapted from Harris, CC, Hamblen, DP, and Francis, JE, Oak Ridge National Laboratory: Basic Principles of Scintillation Counting for Medical Investigators. In: Ross, DA, ed., *Medical Gamma-Ray Spectrometry*, Oak Ridge TN, US Atomic Energy Commn, Division of Technical Information, December 10, 1959.)

Compton valley reflects the sum energy of multiple Compton electrons generated by a single photon (see Fig. 5-8).

Iodine Escape Peak

The photoelectric effect indirectly contributes to the existence of a small peak below the photopeak. When an incoming photon is absorbed as a result of a photoelectric interaction in the sodium iodide crystal, a K-shell electron is usually ejected. Because the K-shell vacancy is then filled by an L-shell electron (see Chapter 3) a 28-keV x-ray (the difference in binding energies between the L and K shells of iodine) is emitted. If this x-ray escapes detection, the total energy absorbed from the original photon is diminished by 28 keV, and a new, small peak is created 28 keV below the photopeak (Fig. 5-9). This is called the **iodine escape peak**. It can only be seen as separate from the photopeak at relatively low energies where the spread of the photopeak is relatively small; at higher photopeak energies, the iodine escape peak is obscured by the spread of the photopeak.

Annihilation Peak

If an entering photon is energetic enough (>1.02 MeV), it may be absorbed near the nucleus of an atom, creating a positron and an electron. This process is called **pair production**. The positron (β^+) will undergo annihilation with an electron, producing two 511-keV photons. In the same reaction, a new photon will be emitted with an energy 1.02 MeV less than that of the incident photon. If the energy of all three photons is detected by the crystal, the total absorbed energy will be equal to the original energy of the incident photon and will contribute to the photopeak. If, however, one 511-keV photon escapes the detector, the sum will be reduced by 511 keV; if both photons escape, the sum will be reduced by 1.02 MeV. The resulting peaks are called the **single escape** and **double escape annihilation peaks**, respectively (Fig. 5-10).

Coincidence Peak

Some nuclides emit two or more photons. Most often, each of these produces its own

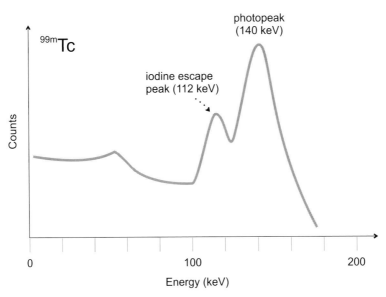

Figure 5-9 Iodine escape peak. (Adapted from Harris, CC, Hamblen, DP, and Francis, JE, Oak Ridge National Laboratory: Basic Principles of Scintillation Counting for Medical Investigators. In: Ross, DA, ed., *Medical Gamma-Ray Spectrometry*, Oak Ridge TN, US Atomic Energy Commn, Division of Technical Information, December 10, 1959.)

characteristic photopeak in the spectrum. However, if two photons impact the crystal simultaneously, the detector will record only a single event with an energy equal to the sum of the two photon energies. This result is the so-called **coincidence peak** (Fig. 5-11).

Effect of Surrounding Matter on the Energy Spectrum

So far we have considered the appearance of the energy spectrum for a point source in air. However, the daily practice of nuclear medicine rarely involves imaging point sources in air with

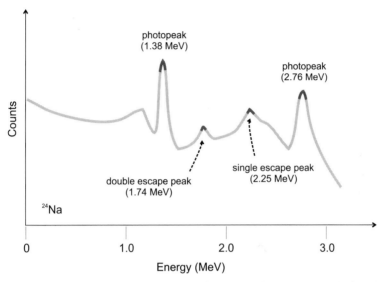

Figure 5-10 Annihilation peaks. (Adapted from Ross, DA, Harris, CC, *The Measurement of Clinical Radioactivity*, Oak Ridge TN, Oak Ridge National Laboratory (ORNL-4153), February 1968.)

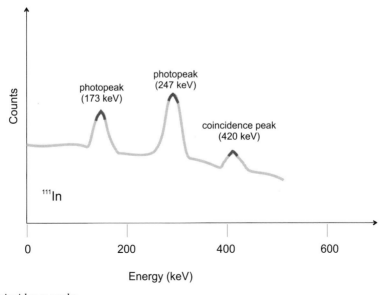

Figure 5-11 Coincidence peaks.

unshielded NaI(Tl) crystals. The lead used to shield the crystal and the water and tissue surrounding the radioactive source alter the shape of the energy spectrum.

Backscatter Peak

When lead shielding surrounds the crystal, photons may exit the crystal without detection only to be deflected 180° off the lead back into the crystal. These photons account for the **backscatter peak**. Their energy is equal to

$E_{backscattered\ photon}$

$$= 256 \times E_{incident\ photon} / (256 + E_{incident\ photon})$$
(Eq. 5-2)

Backscatter peaks are only evident when the incident energy is great enough to contribute a significant degree of Compton scattering in lead (approximately 200 keV).

Characteristic Lead X-Ray Peak

A photoelectric interaction in the lead shield will result in ejection of a K-shell electron along with the prompt emission of an x-ray. This x-ray energy is 72 keV and is equal to the difference between the binding energies of the L and K shells. The emitted x-ray can be detected by the crystal and contributes to the **characteristic lead x-ray peak** at 72 keV.

Additional Compton Scattering from the Medium Surrounding the Source

Many of the photons emitted by a source will undergo Compton scattering in surrounding water or tissue. As a consequence, there will be fewer counts in the photopeak. The Compton photons may be seen by the detector and add to the counts below the photopeak. The effects of water on the energy spectrum of Cr-51 can be seen in Figure 5-12.

Characteristics of Scintillation Detectors

Energy Resolution

Because t-he peaks of an energy spectrum are not sharp spikes but are statistically broadened,

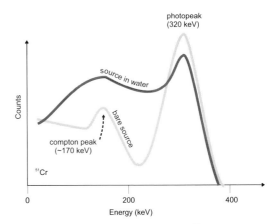

Figure 5-12 The effect of water on the ^{51}Cr energy spectrum. (Adapted from Harris, CC, Hamblen, DP, and Francis, JE. Oak Ridge National Laboratory: Basic Principles of Scintillation Counting for Medical Investigators. In: Ross, DA, ed., *Medical Gamma-Ray Spectrometry*, Oak Ridge TN, US Atomic Energy Commn, Division of Technical Information, December 10, 1959.)

the detector may not be able to separate peaks produced by photons of similar energy. The distance expressed in units of energy between the closest peaks that the detector can distinguish defines the energy resolution of the detector.

Decay Time

Although scintillation detectors will convert incoming radiation to light *whenever* the radiation enters the crystal, this conversion requires about 230 nanosecs for a NaI(Tl) crystal. This interval is called the **decay time** of the crystal. A second gamma photon entering the crystal during this time *will* add to the total output of the light pulse, but it cannot be distinguished as a separate photon. The energy measured by the analyzer will be the sum of the energies of the two gamma photons.

Efficiency

The fundamental meaning of **efficiency** is the *amount observed* as a fraction of the *amount expected*. The **overall efficiency** of a detector can be considered in terms of **geometric** and **intrinsic efficiency**.

Overall Efficiency

The ratio of the number of counts actually registered by the detector to the number of decays generated by a source in a given period of time is the **overall efficiency**. For a well detector, this is simply the number of counts recorded divided by the number of decays in the specimen. For a camera, it is the ratio of the number of counts within the image of the target organ, the so-called region of interest (ROI), to the number of decays of the radionuclide present in the organ. Overall efficiency can be measured, or at least reasonably approximated, by counting a sample container filled with a known quantity of radioactivity.

Using this method, the efficiency is given by the ratio of the number of counts recorded (if using a well detector) or seen in the ROI (if using a camera) to the number of actual decays. Except for attenuation within the sample or its container, the number of decays in the specimen will be 10^6 decays per megabecquerel (MBq) or, by definition, 3.7×10^4 decays per microcurie (μCi).

As a practical matter, the efficiency of modern detectors is not a consideration in the daily work of the nuclear medicine department. Not only are relatively high count rates available from patients and specimens, but most quantitative work is based on the comparison of the sample (such as a urine sample) to a standard (such as the standard used for a Schillings test) count rate using the same device. Because the efficiency is the same for both standard and sample, the exact value of the efficiency is inconsequential. The detector efficiency is more important when measuring the radioactivity in a wipe test. **Wipe tests** are performed by wiping a small absorbent pad over a countertop or other surface and measuring the pad for any radioactive contamination. Here it is necessary to determine whether the activity measured on an absolute scale is below a prescribed limit, for example, less than 37 Bq/cm^2 (1 nCi/cm^2).

Geometric Efficiency

The size and shape of the detector relative to the organ being imaged determines how much of the radiation emanating from the organ is actually "seen" by the face of the detector. Because radiation emanates in all directions, a detector receives only a fraction of the total. The fraction of radiation striking, or "seen" by, the face of the detector obviously will be less for a smaller crystal than larger ones and will further decrease the farther away the detector is placed from the source.

Geometric efficiency is the ratio of the number of photons striking the face of the detector to the number of photons emitted by the target organ, assuming no significant losses in the air between patient and detector. Because the measurement of the number of photons actually striking the face of the detector may be difficult, it is usually satisfactory to infer the geometric efficiency from the physical measurement of the area of the detector face and the distance between it and the target.

Intrinsic Efficiency

Intrinsic efficiency, a component of overall efficiency, is the ratio of the number of counts recorded by the system to the number of photons striking the face of the detector. The **intrinsic efficiency** can be calculated most easily if a known source emitting a known number of photons is placed directly against the face of the detector. In this arrangement, any effect of geometric factors is usually small enough to be ignored.

Types of Crystal Scintillation Detectors

Some commonly used scintillation detectors are described below. All contain a sodium iodide crystal coupled to a single photomultiplier tube. The output of the photomultiplier tube is then routed through the preamplifier, amplifier, and pulse-height analyzer, as described above.

Thyroid Probe

The **thyroid probe** (Fig. 5-13) is a crystal scintillation detector for measuring the radioactivity

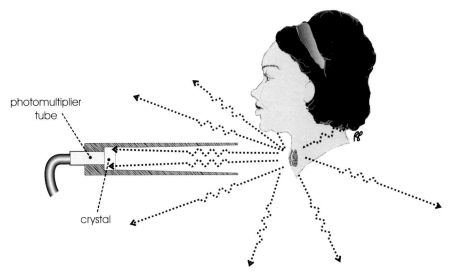

Figure 5-13 Thyroid probe. Shielded crystal scintillation detector as used for measuring thyroid uptake.

in a patient's thyroid gland. The detector is supported on a stand by an adjustable arm that permits placement of the detector against the patient's neck. A cylindrical or slightly conical extension of the shielding around the crystal limits the field of view to the region of the thyroid gland. For measuring ^{123}I or ^{99m}Tc uptake in the gland, the sodium iodide scintillation crystal need be only one-half centimeter thick. Thicker crystals, up to 2-cm thick, are more efficient for higher energy radionuclides, such as ^{131}I.

Well Counter

The **well counter** (Fig. 5-14) is a shielded crystal with a hole, the "well" drilled in the center to hold a specimen in a test tube or vial. In this arrangement, the specimen is surrounded by the crystal so that only a small fraction of the radiation escapes through the opening of the well. The size and thickness of the crystal are selected for efficient capture of photons. For radionuclides with low energy emissions, typical crystals are cylinders of 2 cm to 3 cm in diameter. Two or three times larger crystals are more efficient for higher energies such as those of ^{131}I or ^{59}Fe.

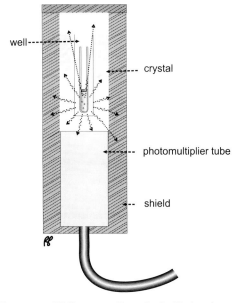

Figure 5-14 Well counter. Crystal scintillation detector constructed with an internal well for counting samples in vitro.

Dosimeters and Area Monitors

Scintillation detectors are available for dosimeters and handheld survey meters for the purpose of monitoring or searching for a source. Others are placed more permanently for area monitoring. In one commercial design, detectors are arranged in a door-like frame through which personnel must pass before leaving a restricted area of the laboratory. The device is often equipped with an audible alarm to signal when the radiation is above an acceptable level.

Appendix: Calculation of the Maximum Compton Electron Energy Corresponding to the Compton Peak (EDGE)

The calculation of the maximum Compton electron energy is fairly straightforward, beginning with the following formula [1] for the energy of the scattered photon, which is assumed as given:

$$E_{\text{scattered photon}} = E_{\text{incident photon}}$$
$$\times [1 + (E_{\text{incident photon}}/0.511)(1 - \cos(\theta))]^{-1}.$$
$$(Eq.\ 5\text{-}3)$$

In addition, it is known that as a result of Compton scattering the Compton electron is ejected from the atom with an energy equal to the difference in the energy between the incident photon and the scattered photon or

$$E_{\text{compton electron}}$$
$$= E_{\text{incident photon}} - E_{\text{scattered photon}} \quad (Eq.\ 5\text{-}4)$$

At *minimal* deflection, $\cos(\theta)$ becomes $\cos(0°)$, which equals 1.0; the above equation then becomes

$$E_{\text{scattered photon}} = E_{\text{incident photon}}$$
$$\times ([1 + (E_{\text{incident photon}}/0.511)(1 - 1)])^{-1}$$
$$= E_{\text{incident photon}}/[1 + 0]$$
$$= E_{\text{incident photon}} \quad (Eq.\ 5\text{-}5)$$

Subsituting the result of Equation 5-5 into Equation 5-4 yields

$$E_{\text{compton electron}} = 0$$

At *maximal* deflection, $\cos(\theta)$ becomes $\cos(180°)$, which equals -1.0; and the equation becomes

$$E_{\text{scattered photon}} = E_{\text{incident photon}}$$
$$\times ([1 + (E_{\text{incident photon}}/0.511)(1 - (-1))])^{-1}$$
$$= E_{\text{incident photon}}$$
$$\times ([1 + (E_{\text{incident photon}}/0.511)(2)])^{-1}$$
$$= E_{\text{incident photon}}/[1 + (E_{\text{incident photon}}/0.256)]$$
$$= 0.256 \times E_{\text{incident photon}}$$
$$\times [(0.256 + E_{\text{incident photon}})]^{-1} \quad (Eq.\ 5\text{-}6)$$

Substituting the result of Equation 5-6 into Equation 5-4 yields

$$E_{\text{compton photon}}$$
$$= E_{\text{incident photon}} - [0.256 \times E_{\text{incident photon}}$$
$$\times [(0.256 + E_{\text{incident photon}})]^{-1}]$$

$$E_{\text{compton photon}} = E_{\text{incident photon}}^2$$
$$\times [(E_{\text{incident photon}} + 0.256)]^{-1} \quad (Eq.\ 5\text{-}7)$$

Reference

1 Cherry, SR, Sorenson, JA, and Phelps, ME, *Physics in Nuclear Medicine*, 3rd Edition, Saunders, Philadelphia, 2003, Chapter 6, Interaction of Radiation with Matter, p. 77.

Test Yourself

1 Connect the following events with the corresponding components of a scintillation detector:
   (a) Photoelectrons are released in response to light photons.
   (b) Light photons are released in response to gamma photons.
   (c) The output of the PMT is amplified to a more readily detectable level of current.
   (d) The multiplying action of successive steps yields a very large number of electrons in response to each photoelectron.
      (i) Photomultiplier tube (PMT)
      (ii) Preamplifier
      (iii) Crystal
      (iv) Photocathode surface of PMT.

2 Name the following features that can be seen in the energy spectrum obtained from a sodium iodide crystal detector:

(a) can be seen at 28 keV below photopeak

(b) energy of photon source

(c) can be seen at 511 keV and 1020 keV below photopeak of high-energy photon

(d) peak near 417 keV emitted by an ^{111}In source

(e) maximum Compton electron energy

(f) 72 keV

  (i) Photopeak

  (ii) Compton peak or edge

  (iii) Annihilation peaks

  (iv) Iodine escape peak

  (v) Coincidence peak

  (vi) Lead x-ray peak

  (vii) Backscatter peak.

3 True or false: A detector that can distinguish separate photopeaks at 200 and 205 keV has a better energy resolution than a detector that can only distinguish photopeaks as close as 200 KeV and 250 keV.

4 Match each of the listed terms (i) through (iii) below to one of the following definitions:

(a) Electrical output from the preamplifiers and amplifiers connected to the PMTs in response to photon interaction in the crystal

(b) Range of acceptable photon energies

(c) Used to select Z pulses within the upper and lower limits of the energy window

  (i) Pulse height analyzer (PHA)

  (ii) Energy window

  (iii) Z pulse.

CHAPTER 6

6 Imaging instrumentation

Theory and Structure

The daily workload in a nuclear medicine department consists of "functional" imaging of organs including thyroid, brain, heart, liver, and kidney. This is accomplished using a large scintillation device. In the 1950s, Dr. Harold Anger developed the basic design of the modern nuclear medicine camera. The **Anger camera** was a significant improvement over its predecessor, the rectilinear scanner. The components of the Anger camera are depicted in Figure 6-1.

Components of the Imaging System

The following components of the imaging system are described in the order they are encountered by the gamma ray photons emitted from the patient's body.

Collimators

A **collimator** restricts the rays from the source so that each point in the image corresponds to a unique point in the source. Collimators are composed of thousands of precisely aligned **holes** (channels), which are formed by either casting hot lead or folding lead foil. They are usu-ally depicted in cross section (Fig. 6-2). Nuclides emit gamma ray photons in all directions. The collimator allows only those photons traveling directly along the long axis of each hole to reach the crystal. Photons emitted in any other direction are absorbed by the **septa** between the holes (Fig. 6-3). Without a collimator in front of the crystal, the image would be indistinct (Fig. 6-4).

There are several types of collimators designed to channel photons of different energies. By appropriate choice of collimator it is possible to magnify or minify images and to select between imaging quality and imaging speed.

Parallel-Hole Collimators

Low-energy all-purpose collimators (LEAP): These collimators have relatively large holes which allow the passage of many of the photons emanating from the patient. As such they have relatively high sensitivity at the expense of resolution. Because the holes are larger, photons arising from a larger region of the source are accepted. As a result, image resolution is decreased (see Figs 6-5 and 6-7 and associated text). The sensitivity of one such collimator has been calculated at approximately 500,000 cpm for a 1-μCi source, and the resolution is 1.0 cm at 10 cm from the surface of the collimator (Nuclear Fields Precision Microcast Collimators, Nuclear Fields B.V.,

The Netherlands; 140-keV ^{99m}Tc source). These collimators are useful for imaging low-energy photons such as those from ^{201}Tl for which thick septa are not necessary. In addition, because of their moderately high sensitivity (resulting from thinner septa and bigger holes) they are advantageous for images of short duration such as the sequential one-per-second images for a renal flow study.

DISTINCTION

Nuclear medicine imaging systems are called cameras; just as standard cameras image light, they image photons generated by nuclides. Unlike an x-ray machine that creates the photons (x-rays) that penetrate the body and detects those that emerge on the other side of the patient, nuclear medicine camera utilize gamma photons from radio-pharmaceuticals within the patient.

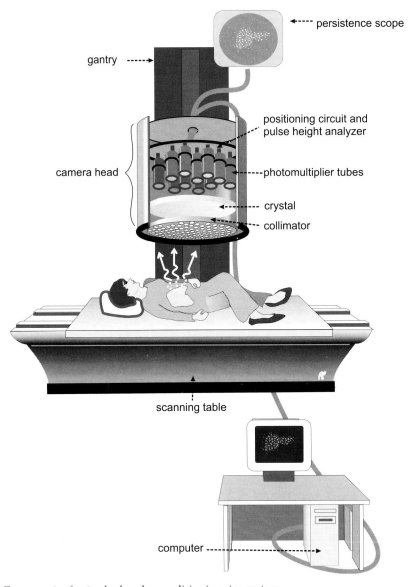

Figure 6-1 Components of a standard nuclear medicine imaging system.

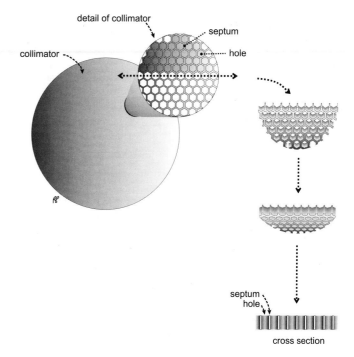

Figure 6-2 Collimator detail.

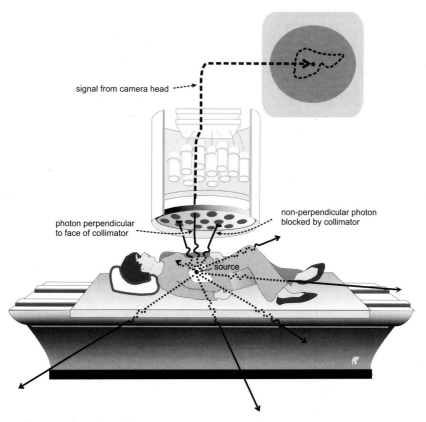

Figure 6-3 A collimator selects photons perpendicular to the plane of the collimator face.

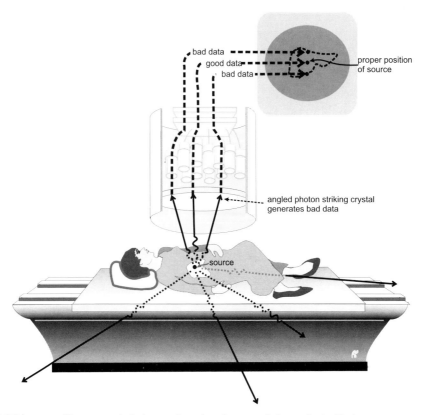

Figure 6-4 Without a collimator, angled photons introduce improperly located scintillations.

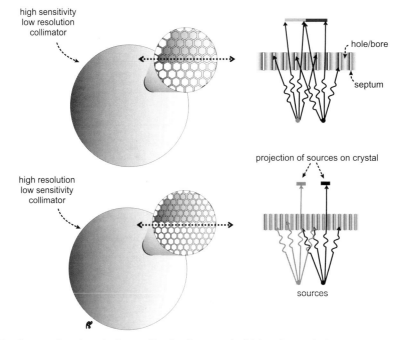

Figure 6-5 For the same bore length, the smaller the diameter the higher the resolution.

High-resolution collimators: These collimators have higher resolution images than the LEAP collimators. They have more holes that are both smaller in diameter and longer in length. The calculated sensitivity of a representative high-resolution collimator is approximately 185,000 cpm for a 0.037 MBq (1-μCi) source, and its nominal resolution is 0.65 cm at 10 cm from the face of the collimator (Nuclear Fields B.V.).

To compare the performance of a LEAP with a high-resolution collimator, let us look at the photons from two radioactive points in a liver. The photons from each of the points are emitted in all directions, but the detector can "see" only those photons that pass through the holes of the collimator. The relatively large hole in the LEAP collimator will also admit photons scattered at relatively large angles from the direct line between the liver and the crystal. This lowers the resolution because the angled photons have the effect of merging the images of two closely adjacent points (Fig. 6-5). At the same time, the larger holes and correspondingly thinner septa give the LEAP a higher sensitivity by admitting a higher percentage of the photons. The high-resolution collimator, on the other hand, admits photons from a smaller fraction of the organ, because more of its face is blocked by septa. In a reciprocal way, the narrower (see Fig. 6-5) and/or longer (Fig. 6-6) bore of its holes better collimate those photons that do enter the collimator. As a result, relatively closely spaced details in the source are more likely to appear clearly separated in the image.

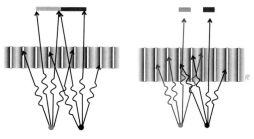

Figure 6-6 For the same hole diameter, the longer the bore the higher the resolution.

SENSITIVITY

Refers to the ability of the imaging camera to detect the photons generated by the nuclides. A **low-sensitivity** system detects a smaller number of the generated photons; a **high-sensitivity** system detects a greater number. For a given crystal, increasing the area occupied by holes as opposed to septa increases sensitivity. As a practical matter, this means larger holes and thinner septa. Another technique to improve sensitivity is to increase the thickness or area of the crystal.

RESOLUTION

Refers to the ability of the camera to distinguish between adjacent points in the organ, permitting detection or finer detail. The higher the resolution the closer are the points that can be distinguished. Using holes of longer and/or narrower bore increases resolution. A collimator's resolution may be expressed as its angle of acceptance. Only photons falling within the **angle of acceptance** can reach the crystal (Fig. 6-7).

There is a trade-off between sensitivity and resolution. A high-sensitivity system has a relatively low resolution. To understand this, remember that although bigger holes admit more photons (higher sensitivity), they reduce the resolution (larger angle of acceptance).

High- and medium-energy collimators: Low-energy collimators are not adequate for the higher-energy photons of nuclides such as [67]Gallium (it emits 394-keV, 300-keV, and 185-keV photons, in addition to its low-energy 93-keV photon), [131]Iodine (376 keV), [111]Indium (245 keV and 173 keV), nor for positron emitters such as [18]Fluorine (511-keV annihilation). The photons of these nuclides can penetrate the thinner septa of both the LEAP and the high-resolution collimators, resulting in poorer resolution. High-energy collimators with thicker septa (Fig. 6-8) to reduce septal penetration are used, but thicker septa also mean smaller holes and, consequently, lower sensitivity. High-energy collimators are useful for [131]Iodine.

high sensitivity collimator

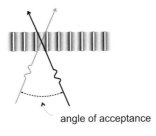

angle of acceptance

high resolution: narrower holes

high resolution: longer bores

Figure 6-7 Angle of acceptance.

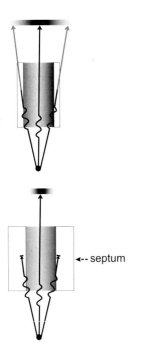

←-- septum

Figure 6-8 Thicker septa are used to block high- and medium-energy photons.

Medium-energy collimators have characteristics between those of low- and high-energy collimators. They can be used to image photons emitted by ^{67}Gallium and ^{111}Indium. The terms high-, medium-, and low-energy are not rigidly defined, and usage may vary from institution to institution.

Slant-hole collimators: These are parallel-hole collimators with holes directed at an angle to the surface of the collimator. The slant-hole collimator provides an oblique view for better visualization of an organ that would otherwise be obscured by an overlying structure while permitting the face of the collimator to remain close to the body surface (Fig. 6-9).

MEMORY HINT

Remembering the difference between diverging and converging collimators is easier if you realize that in the former the holes diverge (or spread out) toward the patient, in the latter they converge (come together) toward the patient.

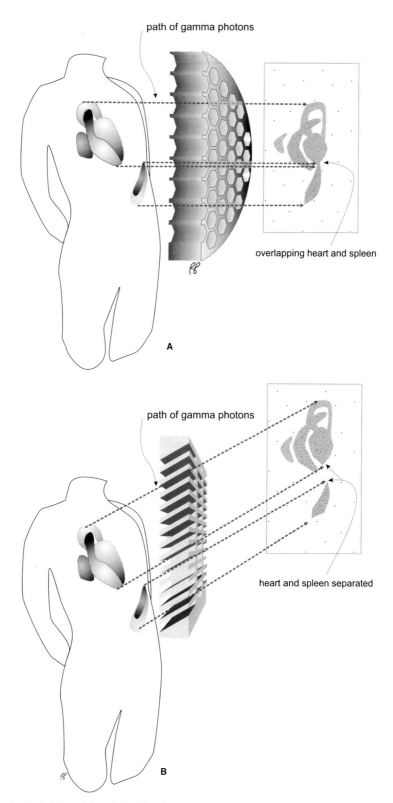

path of gamma photons

overlapping heart and spleen

A

path of gamma photons

heart and spleen separated

B

Figure 6-9 Parallel-hole (A) and slant-hole (B) collimators.

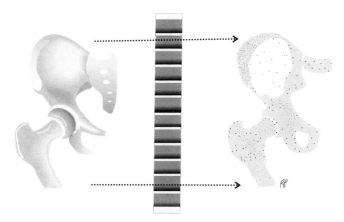

Figure 6-10 Parallel-hole collimator.

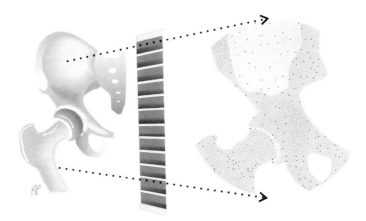

Figure 6-11 Converging collimator.

Nonparallel-Hole Collimators

Nonparallel-hole collimators provide a wider or a narrower field of view. The cone-like pattern of holes allows these collimators to enlarge or reduce the size of the image.

Converging and diverging collimators: A parallel-hole collimator and its image are shown in Figure 6-10. At least for this simple example, the organ and the image are the same size. In a converging collimator, however, the holes are not parallel but are angled inward, toward the organ, as shown in Figure 6-11. Consequently, the organ appears larger at the face of the crystal.

Diverging collimators achieve a wider field of view by angling the holes the opposite way, outward toward the organ. This is used most often on a camera with a small crystal, such as a portable camera. Using a diverging collimator a large organ such as the lung can be captured on the face of a smaller crystal (Fig. 6-12).

Pinhole collimators: These have a single hole—the pinhole—usually 2 mm to 4 mm in diameter. Like a camera lens, the image is projected upside down and reversed right to left at the crystal (Fig. 6-13). It is usually corrected electronically on the viewing screen. A pinhole collimator

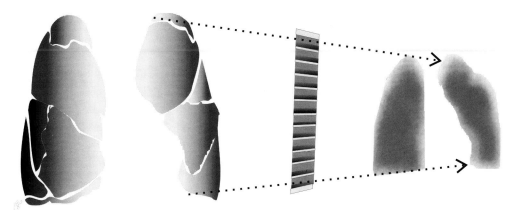

Figure 6-12 Diverging collimator.

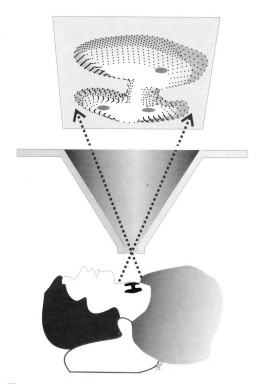

Figure 6-13 Pinhole collimator.

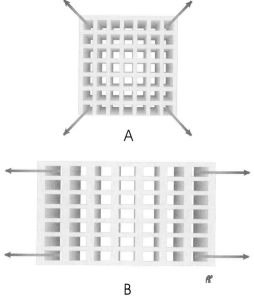

Figure 6-14 (A) In a converging collimator, holes converge toward the patient in both planes and uniformly enlarge an image. (B) In a fan-beam collimator, holes converge in one plane but are parallel in the other. Arrows demonstrate the paths of photons coming from the patient.

generates magnified images of a small organ like the thyroid or a joint.

Fan-beam collimators: These are a cross between a converging and a parallel-hole collimator. They are designed for use on cameras with rectangular heads when imaging smaller organs such as the brain and heart. When viewed from one direction (along the short dimension of the rectangle), the holes are parallel. When viewed from the other direction (along the long dimension of the rectangle), the holes converge (Fig. 6-14). This arrangement allows the data from the patient to be spread to better fill the surface of the crystal.

Camera Head

The camera head contains the crystal, photomultiplier tubes, and associated electronics (see Fig. 6-1). The **head housing** envelopes and shields these internal components. Typically, it includes a thin layer of lead. A **gantry** supports the heavy camera head.

Crystals, Photomultiplier Tubes, Preamplifiers, and Amplifiers

The crystal for an imaging camera is a large slab of thallium-"doped" NaI crystal similar to that used for the scintillation probes described in Chapter 5. It should be noted that the thickness of a crystal affects its resolution as well as its sensitivity. Although thicker crystals have higher sensitivity, the resolution is lower because gamma rays may be absorbed farther from the point at which they entered the crystal (Fig. 6-15).

Sixty or more photomultiplier tubes may be attached to the back surface of the crystal using light-conductive jelly. Each of these function very much like the single photomultiplier tube described in Chapter 5. The preamplifiers and amplifiers are also discussed there.

Positioning Circuit

The amount of light received by a photomultiplier tube (PMT) is related to the proximity of the tube to the site of interaction of the gamma ray in the crystal. The photomultiplier tube closest to the site of interaction receives the greatest number of photons and generates the greatest output pulse; the tube farthest from the nuclide source receives the fewest light photons and generates the smallest pulse. Although an image can be composed solely of the points corresponding to the photomultiplier tube with the highest output at each photon interaction, the number of resolvable points is then limited to the total number of PMT tubes (up to 128 per camera).

A **positioning circuit** improves resolution by factoring in the output from adjacent tubes (Fig. 6-16). The positioning circuit uses a "voltage divider" to weigh the output of each tube in relation to its position on the face of the crystal. A simple voltage divider can be constructed from an electrical resistor with an intermediate tap. The full voltage is applied across the ends of the resistor. The voltage at the tap is

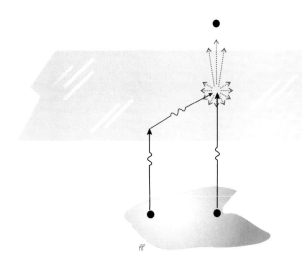

Figure 6-15 Scattering of photons in a thicker crystal reduces resolution.

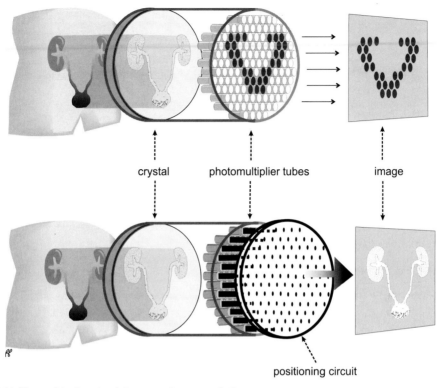

Figure 6-16 The positioning circuit improves image resolution.

proportional to its placement on the resistor (Fig. 6-17).

RESISTOR

A basic understanding of resistors will help to better understand how a voltage divider works. A simplified circuit is shown in the upper left panel of Figure 6-17. The 1-V battery causes electrons to flow through the wire. A resistor is introduced into the circuit (upper right panel) and may be thought of as resisting the flow of electrons, converting part of their kinetic energy into heat. The voltage drop across the full length of the resistor is 1 V, equal to the voltage supplied by the battery. The loss of kinetic energy and the voltage drop are proportional to the distance the electrons travel across the resistor. A voltmeter placed one half of the distance along the resistor will register 0.5 V (bottom left panel), while a detector placed three quarters of the way along the resistor will register 0.75 V (bottom right panel).

The following is a rudimentary explanation of a positioning circuit. The output of each of the preamplifiers attached to each PMT is connected to four directional terminals: X^+, X^-, Y^+, and Y^-. The size of the current pulse reaching each of the four terminals from a preamplifier is dependent on the proximity of the PMT to each terminal. Figure 6-18 depicts how resistors can be used to weight (increase or decrease) the size of the current pulse reaching the terminals based on the distance between the PMT and each terminal. The upper left panel Figure 6-18 is a simplified diagram of 20 PMT tubes, each of which is connected through a resistor to the X^- terminal. Following a photon interaction in the crystal the output from each PMT tube is weighted in proportion to its distance from the X^- terminal. The output for the X^- terminal is the sum of these weighted outputs from the 20 PMTs.

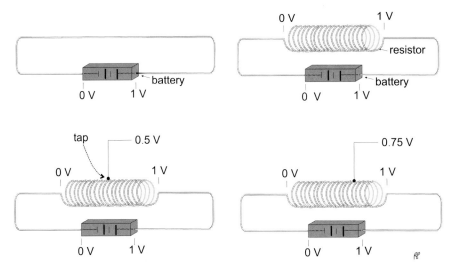

Figure 6-17 Using a resistor as a voltage divider.

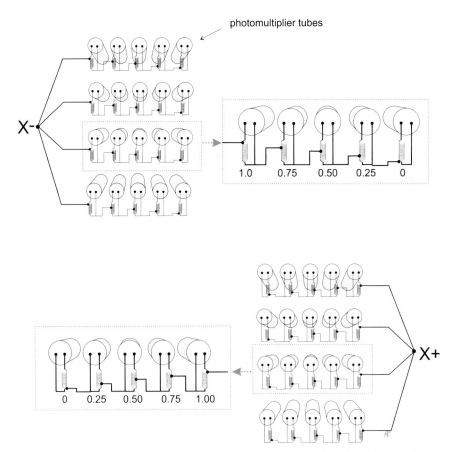

Figure 6-18 The X-axis positioning circuit divides the PMT outputs between the X^+ and X^- terminals.

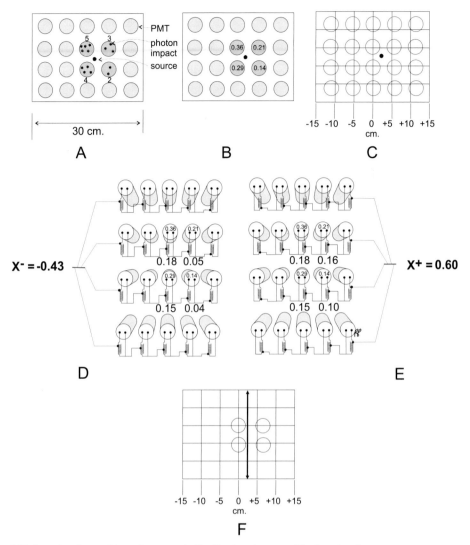

Figure 6-19 Locating the position of a source in the X-axis using a positioning circuit.

Referring to the upper right panel of Figure 6-18, we see that the first tube is closest to the X$^+$ terminal; therefore, the full voltage output from this tube goes to the X$^+$ terminal. The next PMT is one fourth of the distance away, so we discard a quarter of the output and retain the remaining three fourths. Finally, for the last tube all of the X$^+$ output is discarded.

The lower panels of Figure 6-18 depict the connection of the same PMTs to the X$^+$ terminal. In this case, the right-hand tube is closest to the terminal so the full output voltage goes to the

X$^+$ terminal, while all of the output of the left-hand tube is discarded. The Y$^-$ and Y$^+$ terminals, which are not illustrated, are connected to the PMTs in a similar fashion.

Figure 6-19 shows the steps involved in "calculating" the location of the site of interaction in the X direction as a sum of the pulses reaching the X$^-$ and X$^+$ terminals. In this simplified example, a 30-cm crystal is attached to 20 PMTs (separated from each other only for the purposes of illustration). When a gamma photon interacts with the crystal at "source," four PMTs receive light photons (Fig. 6-19A). The upper left of these four tubes receives the greatest number of photons (36% of the total) since it is closest to the source; the other three tubes receive 29%, 21%, and 14%, respectively (Fig. 6-19B). The grid sketched in Figure 6-19C shows that the initial interaction in the crystal is approximately 2.5 cm from the crystal center in the X^+ direction. In the next step of the process, the signal from each of the four tubes is multiplied by its weighting factor as given in Figure 6-18. For example, the output of the upper left of the four tubes, which received 36% of the total signal, is multiplied by a weighting factor of 0.5, yielding a weighted output of 0.18 (Fig. 6-19D, E). Finally, the sum of the X^- and X^+ terminal outputs are combined

$$X^+ + X^- = X = 0.60 - 0.43 = 0.17$$

The positioning circuit places the source at

$$0.17 \times 15 \text{ cm} = 2.55 \text{ cm}$$

from the center of the crystal in the X^+ direction (Fig. 6-19F). The output of the Y^- and Y^+ terminals are processed in the same way.

Pulse-Height Analyzer

The function of the pulse-height analyzer is discussed in Chapter 5. Following each gamma photon interaction in the crystal, the sum of output signals from the X^-, X^+, Y^-, and Y^+ is proportional to the energy of the gamma photon striking the crystal. This summed output is called the **Z-pulse**. The pulse-height analyzer accepts only those Z-pulses that correspond to the gamma energy of interest. Each accepted Z-pulse with its location (as determined by the positioning circuit) is stored in the computer.

Persistence Scope

In the past, the **persistence scope** was made with a phosphor that fades very slowly to provide a slowly changing view of the x,y location of the Z-pulses that are accepted (each of which corresponds to the location of a gamma ray). At present, the same result is achieved with much greater flexibility by storing the image in computer memory and using it to update the image on the screen slowly or rapidly. This allows the person who is acquiring the scan to adjust the position of the patient prior to recording the final image on film or computer.

Computers

Nuclear medicine computers are used for the acquisition, storage, and processing of data. The image data are stored in digital form (**digitized**) as follows: For each Z-pulse that is accepted by the pulse-height analyzer, one count is added to the storage location that corresponds to its x,y location determined by the positioning circuit. The data storage (area of the coumputer's memory) can be visualized as a **matrix**, a kind of two-dimensional checkerboard. Each position within the matrix corresponds to a **pixel** within the image and is assigned a unique "address" composed of the row and column of its location (Fig. 6-20). Data are digitized by assigning a matrix position to every accepted photon (Fig. 6-21). Matrices are defined by the number of subdivisions along each axis. The operator can select from several matrix configurations of successively finer divisions; 64×64, 128×128, 256×256, and 512×512, or more. (The preceding are all examples of square matrices.) These numbers refer to the number of columns and rows in a square matrix. Notice that the outside dimensions of all matrices are the same size. What varies is the pixel size and hence the total number of pixels. A 64×64 matrix has 4096 pixels; a 128×128 matrix has 16,384 pixels; and so on.

The greater the number of pixels the smaller is each pixel for a given field of view and the better preserved is the resolution of the image (Fig. 6-22). The camera and computer system cannot reliably distinguish between two points that are separated by less than

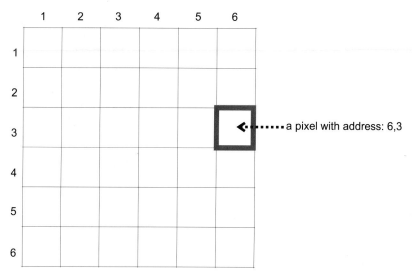

Figure 6-20 Small matrix.

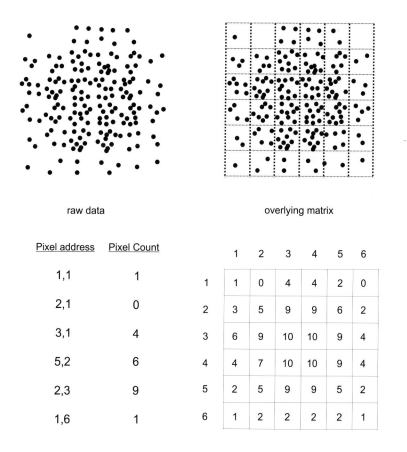

raw data

overlying matrix

Pixel address	Pixel Count
1,1	1
2,1	0
3,1	4
5,2	6
2,3	9
1,6	1

	1	2	3	4	5	6
1	1	0	4	4	2	0
2	3	5	9	9	6	2
3	6	9	10	10	9	4
4	4	7	10	10	9	4
5	2	5	9	9	5	2
6	1	2	2	2	2	1

matrix of pixel counts

Figure 6-21 Storing image data in a matrix.

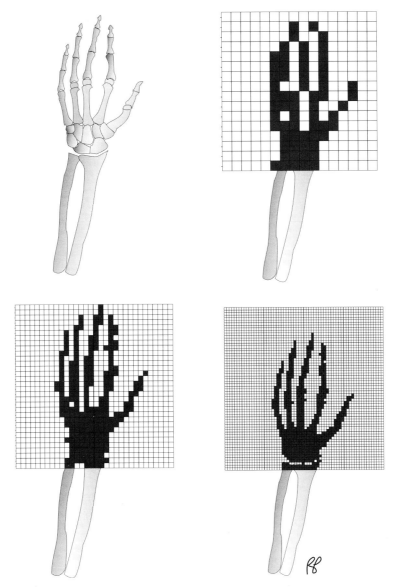

Figure 6-22 Effect of matrix configuration on image resolution.

1 pixel. In Figure 6-23, three "hot spots" are seen in the kidney parenchyma. With the coarser matrix (larger pixel size) depicted in the upper portion of the figure, two of the three spots are within a single pixel. In the finer matrix pictured in the lower portion of the figure, all three points are distinct as there are now several "cold" pixels between the "hot" pixels.

The size of a pixel is inversely related to the dimensions of a matrix for a given field of view. For a 32-cm camera field of view mapped onto a 64 × 64 computer matrix, each pixel will measure 0.5 cm on a side. For the same camera, each pixel of a 128 × 128 matrix will measure 0.25 cm on a side. Similarly, a matrix of 256 × 256 will divide the field of view into pixels measuring 0.125 cm on a side. This means that points that

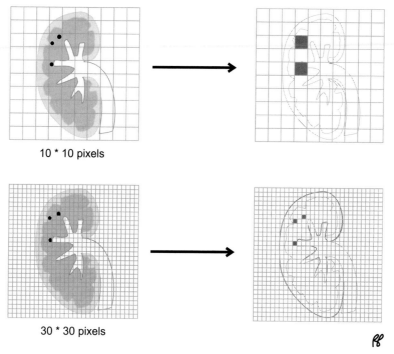

10 * 10 pixels

30 * 30 pixels

Figure 6-23 A matrix cannot resolve points separated by less than 1 pixel.

are closer to each other than 0.5 cm, 0.25 cm, and 0.125 cm, respectively, cannot be separately distinguished.

Of course, the maximum resolution of the image is limited by the resolution of the camera and collimator. Once the data is stored in a matrix, images can be displayed on the computer screen or film.

Planar Imaging

Image Acquisition

The gamma photons emitted by the patient may be acquired in various forms: static, dynamic, gated, or planar tomographic images.

Static Images

Static imaging is used to collect images of different regions of the body or differently angled (oblique) views of a particular region of interest. For example, a bone scan is composed of static images of 12 different regions of the body (three head views, left arm, right arm, anterior chest, posterior chest, and so on) or anterior and posterior whole body images with additional static views of select regions of the skeleton. The whole body images seen in Figure 6-24 were obtained by moving the gantry at a steady rate of 12 cm/min while acquiring the imaging data. Liver scans are generally composed of six angled views of the liver and spleen; thyroid scans may have three or four angled views of the thyroid. In all of the above cases there is very little change in the distribution of nuclide in the organ of interest while the images are being acquired.

Dynamic Images

If the distribution of nuclide in the organ is changing rapidly and it is important to record this change, multiple rapid images of a particular region of interest are acquired. This type of image acquisition is called **dynamic imaging**.

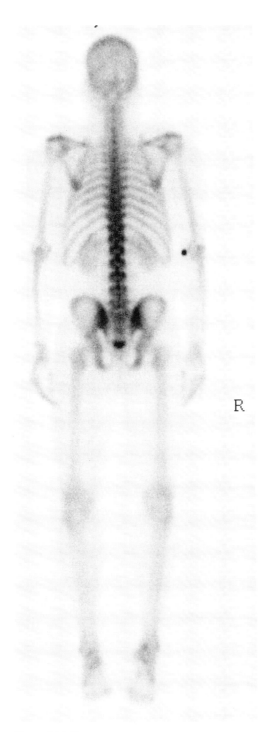

R

Figure 6-24 Bone scan.

It is used, for example, to collect sequential 1-second images (called **frames**) of the vascular flow of nuclide through the kidney. In the example shown in Figure 6-25, these 60 frames are summed or **compressed** into 15 frames of 4 seconds each by adding every four frames together.

In Figure 6-26, sequential anterior abdominal images demonstrate an active bleeding site originating in the colon at the splenic flexure. Radiolabled blood progresses through the descending colon to the rectum. The images were first acquired as 96 frames of 30 seconds-duration. These images were then "compressed" into 16 images of three minutes each. Dynamic imaging can be thought of as a type of video recording to "catch" images of fast action; static images are similar to photographs.

Gated Images

Gated images are a variation of dynamic images. Continuous images are obtained of a moving organ (generally the heart) and data are coordinated with the rate of heart beat (using electrocardiographic leads to keep track of the R–R interval for heart imaging). Gating is used to divide the emission data from the radioactive blood pool in a Gated Blood Pool Study (GBPS) or Multigated Acquisition (MUGA) into "frames" so that wall motion can be evaluated and a left ventricular ejection fraction can be calculated. During gated imaging, each cardiac cycle (the interval between R waves on the electrocardiogram) is divided into frames (Fig. 6-27). The output of the electrocardiogram (EKG attached to the patient) is connected to the nuclear medicine camera/computer. Each R wave will "trigger" the collection of a set of frames. In this simplified example, the camera/computer is programmed to collect eight frames of 0.125 seconds each. This assumes that each heart beat is of one second's duration. Images of the cardiac blood pool collected from approximately 600 cardiac cycles are

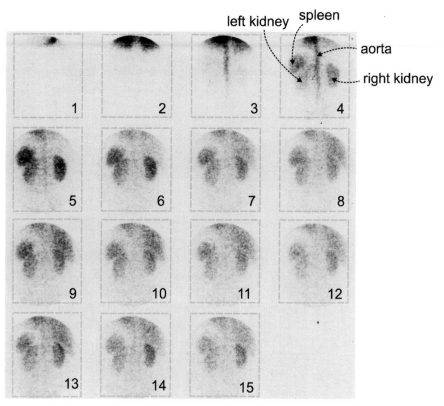

Figure 6-25 Sixty-second renal flow study.

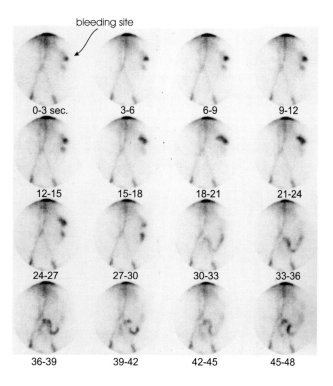

Figure 6-26 Gastrointestinal bleeding scan.

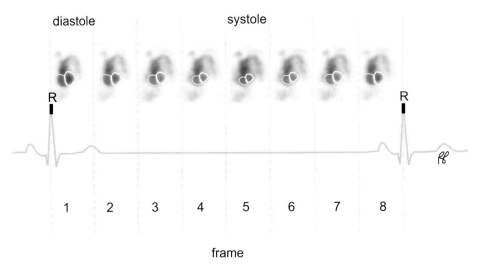

Figure 6-27 Gated blood pool study.

summed to create a single composite gated cycle image.

Test Yourself

1 True or false: A parallel-hole collimator providing both high sensitivity and high resolution is difficult to design principally because the requirements for sensitivity conflict with those for resolution.

2 True or false: A slant-hole collimator is used to visualize an organ that is partially obscured by an overlying structure.

3 True or false: In the converging-hole collimator, the holes angle inward toward the organ to be imaged. In consequence, the organ appears larger at the face of the crystal.

4 How many pixels of data can be stored in a 64×64 image matrix? Choose one.
(a) 128
(b) 256
(c) 512
(d) 1024
(e) 2048
(f) 4096.

5 The usual gamma camera head contains all of the following elements except:
(a) Crystal
(b) Photo-multiplier tubes
(c) Collimator
(d) Positioning circuit
(e) Pulse-height analyzer
(f) Focusing assembly.

7 Single-photon emission computed tomography (SPECT)

Single-photon emission computed tomography (SPECT) cameras acquire multiple planar views of the radioactivity in an organ. The data are then processed mathematically to create cross-sectional views of the organ. SPECT utilizes the single photons emitted by gamma-emitting radionuclides such as ^{99m}Tc, ^{67}Ga, ^{111}In, and ^{123}I. This is in contrast to positron emission tomography (PET), which utilizes the paired 511-keV photons arising from positron annihilation. PET is the subject of Chapter 8.

Equipment

Types of Cameras

The simplest camera design for SPECT imaging is similar to that of a planar camera but with two additional features. First, the SPECT camera is constructed so that the head can rotate either stepwise or continuously about the patient to acquire multiple views (Fig. 7-1). Second, it is equipped with a computer that integrates the multiple images to produce the cross-sectional views of the organ.

The more advanced SPECT camera designs have more than one head or are constructed with a ring of detectors. In the case of the single and multiple head cameras, the heads are mechanically rotated around the patient to obtain

Figure 7-1 SPECT camera.

the multiple projection views (Fig. 7-2). **Ring detectors** have a ring of individual small crystals or a single, donut-shaped crystal that does not rotate (the collimator does rotate) (Fig. 7-3).

Angle of Rotation of Heads

Single-headed cameras must rotate a full 360° to obtain all necessary views of most organs. In contrast, each head of a double-headed camera need rotate only half as far, 180°, and a triple-headed camera only 120° to obtain the same views. The cost of the additional heads must be

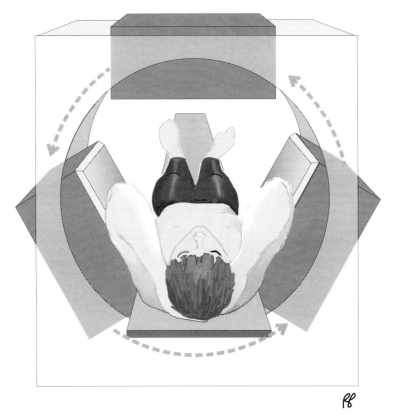

Figure 7-2 Three-headed SPECT camera.

balanced against the benefits of increased speed of acquisition.

Two-Headed Cameras: Fixed and Adjustable

Two-headed cameras can have a fixed, parallel configuration or fixed, perpendicular configuration (Fig. 7-4). Fixed, parallel heads (opposing heads) can be used for simultaneous anterior and posterior planar imaging or can be rotated as a unit for SPECT acquisition. Fixed, perpendicular heads, in an L-shaped unit, are used almost exclusively for cardiac or brain SPECT imaging.

Adjustable heads allow positioning of the heads in different angular configurations. The moveable head can be moved closer to or farther along the ring from the other head so that the two heads are parallel (opposing), perpendicular

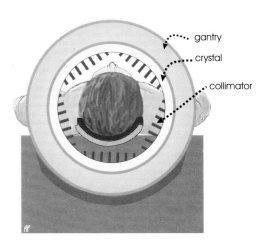

Figure 7-3 SPECT camera with a single donut-shaped crystal.

(L-shaped), or separated by an intermediate angle. Thus the adjustable two-headed camera can be used for planar imaging and for large and small organ tomography.

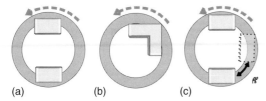

Figure 7-4 Two-headed SPECT camera: (a) fixed, parallel; (b) fixed, perpendicular; (c) adjustable.

TOMOGRAPHY

Tomos is the Greek word for cut or section. Tomography is a name originally used for conventional x-rays that were modified to bring only a single plane through the patient into focus.

Acquisition

The numerous, sequential planar views acquired during tomographic acquisition are called **projection views**. They are little more than an intermediate step toward creating slices. Figure 7-5 shows an entire set of projection views that can be used to construct tomographic images of the liver and spleen. Because of the large number of views, 64 in this case, compared to the five typically used for a conventional liver–spleen scan (Fig. 7-6), these projection views are most useful when displayed as a moderately rapid sequential presentation, the so-called **cine** view. The term cine is used because of its resemblance to movies.

Arc of Acquisition

Tomographic projection views are most often acquired over an arc of 360° or 180°. The 360° arc of rotation of the camera heads is regularly used for most organs. The 180° arc is used for organs that are positioned on one side of the body, such as the heart.

Views of the heart are obtained in a 180° arc extending from the right anterior oblique position to the left posterior oblique position.

The data from this 180° is considered adequate, because photons exiting the body from the right posterior and lateral chest travel through more tissue and suffer greater attenuation than those exiting through the left side (Fig. 7-7).

Number of Projection Tomographic Views

Over a full 360° arc, 64 or 128 tomographic projections are usually collected; similarly 32 or 64 views are generally obtained over a 180° arc.

Collection Times

For a given dose of radiopharmaceutical, better images are generated using the higher count statistics from longer acquisitions. However, patient comfort and cooperation limit imaging times. Acquisition times of 20 to 40 seconds per projection view are standard.

Step-and-Shoot vs. Continuous Acquisition

The standard method for collection of tomographic projection views is called **step-and-shoot acquisition**. In this technique, each projection view is acquired in entirety at each angular **stop** (position). There is a short pause of a few seconds between views to allow for the automatic rotation of the camera head to the next stop. The camera makes a single rotation around the patient. In the example depicted in Figure 7-8, projection views of 20 s duration were obtained every 5.6° for a total of 64 views. The camera paused for 2 s after acquiring each view as the head moved into position for the next view. The total acquisition time was 1408 s. Since 126 s of this total were "consumed" by pauses, there was a total imaging time of 1282 s.

In **continuous acquisition**, data are collected over one or several sequential 360° rotations. There are no pauses; rotation is continuous. In the example depicted in Figure 7-9, the camera rotated a full 360° every 140.8 s. Ten

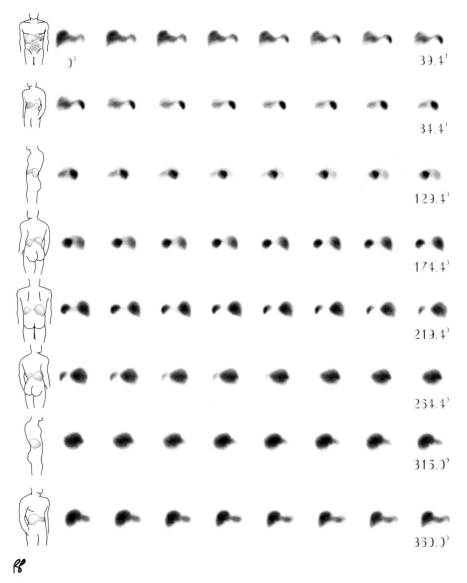

Figure 7-5 Projection views.

such rotations provided 1408 s of imaging time (compared to the 1282 s from the step-and-shoot acquisition).

Circular vs. Elliptical Orbits

Most acquisitions are performed with a **circular orbit**. The camera head is rotated at a fixed distance from the center of the body. Since the body is more nearly elliptical than circular in cross section, the camera does not come as close to the organ as possible over a significant portion of its rotation (left panel of Fig. 7-10). Because image statistics and resolution are better if the camera is as close to an organ as possible, some cameras are designed to rotate in elliptical orbits, which allow the camera head to more closely follow the contour of the body and therefore stay

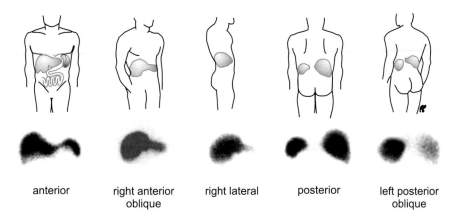

anterior   right anterior   right lateral   posterior   left posterior
oblique   oblique

Figure 7-6 Five-view liver–spleen scan.

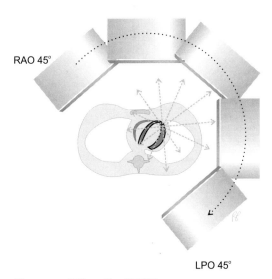

RAO 45°

LPO 45°

Figure 7-7 180° cardiac SPECT.

closer to the organ being imaged (right panel of Fig. 7-10).

Patient Motion and Sinograms

Significant patient motion can cause artifacts or blurring in an image. To detect patient motion, the cine display and/or the **sinogram** (see below) should be reviewed prior to releasing the patient. Small amounts of patient motion can be corrected by automated correction algorithms which shift the projection views

to align the organ of interest. Repeat acquisition is recommended if these algorithms are not successful.

A **sinogram** image is a stack of slices of the acquired projection views from 0° to maximum angle of rotation, either 180° or 360°. Each row of the sinogram image consists of data acquired at a different angle of rotation, but all of the rows in the sinogram come from the same axial(y) position. In other words, there is a separate sinogram image for each slice location along the y-axis (the long axis) of the patient. Figure 7-11A is an illustration of the construction of a sinogram representing a thin slice of the heart obtained from sample projection views from a 180° arc around the patient, Figure 7-11B is the complete sinogram containing all of the projection views. Figure 7-12 shows sinograms taken at the level of the heart, the gallbladder, and the bowel. Figure 7-13 shows discontinuities seen in the sinogram of a patient who moved in the direction of the x-axis (side to side motion) and the y-axis (motion along the long axis) of his body during the acquisition of his images. It is not always possible to distinguish the direction of motion from examination of the sinogram, but the cine images provide this information. Figure 7-14 is an example demonstrating image caused by patient motion in the direction of the y-axis. Correcting motion in the x-axis (created

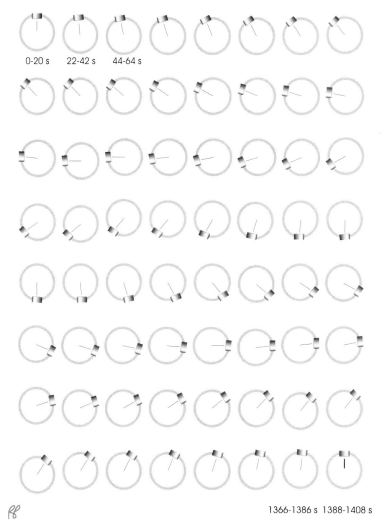

0-20 s   22-42 s   44-64 s

1366-1386 s 1388-1408 s

Figure 7-8 Step-and-shoot acquisition.

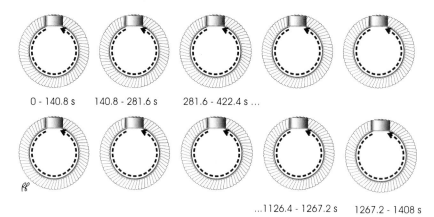

0 - 140.8 s   140.8 - 281.6 s   281.6 - 422.4 s ...

...1126.4 - 1267.2 s   1267.2 - 1408 s

Figure 7-9 Continuous acquisition.

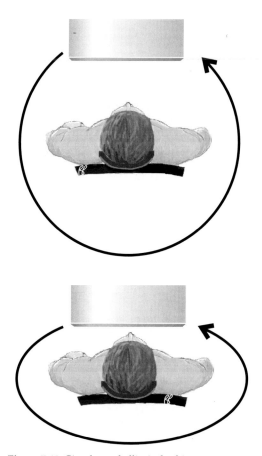

Figure 7-10 Circular and elliptical orbits.

by side to side rotation or shifting in the axial plane) is more difficult than correcting motion in the direction of the y-axis (the long axis of the patient).

Reconstruction

Reconstruction is the process of creating transaxial slices from projection views. There are two basic approaches to creating the transaxial slices. The first, most commonly used, is called **filtered backprojection** and will be covered in depth in the following text. The second technique, **iterative reconstruction,** is mathematically more complex and a brief introduction will be presented at the end of this section.

Filtered Backprojection

The process of backprojection will be introduced, followed by a discussion of filtering.

Backprojection

In **backprojection**, the data acquired by the camera are used to create multiple **transaxial slices**. Figure 7-15 is a representation of this process; the ten projection views are cut into seven bands and shown pulled apart. The bands forming each view are then shown smeared along a radius, like the spokes of a wheel. It is this smearing back toward the center that gives us the term backprojection. The smears of the middle bands (shown in dark grey) are seen in translucent gray.

Figure 7-16 is a representation of data obtained in the acquisition of the projection views of a thin radioactive disk. In this figure, an imaginary grid is placed over the disk (Fig. 7-16A), the disk is imaged (Fig. 7-16B), and the counts for each pixel are recorded (Fig. 7-16C). The counts in each of the cells of a column are summed and stored in an array (Fig. 7-16D). In a similar manner, all of the rows are summed and stored in an array of sums to the right of the matrix (Fig. 7-16E).

During backprojection, the two arrays of projection data are used to recreate the original disk. In the upper panel of Figure 7-17, the upper array is spread or backprojected across the columns of a blank matrix so that each of the values in any single column are identical. The array to the right of the matrix is backprojected across the rows, and these values are added cell by cell to the values of the preceding set (middle panel). If the counts in each pixel are represented by dots (for the ease of illustration each dot represents 5 counts) one begins to see a relatively dense central area that corresponds generally to the size and the location of the original disk (lower panel). The wide bands of dots extending in four directions from this central density are an artifact of the backprojection process; they are residual counts from the backprojection of the arrays.

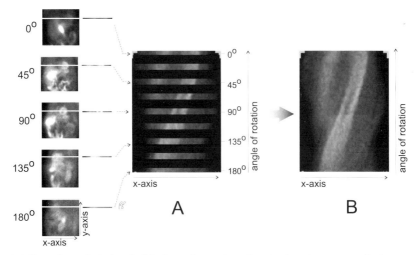

Figure 7-11 (A.) Slices through the level of the heart from selected projection views are stacked to create a sinogram. (B) Complete sinogram.

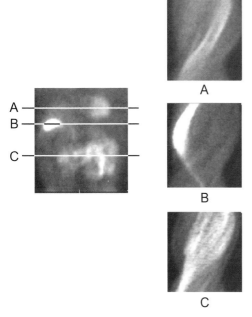

Figure 7-12 Sinograms at selected positions along the long axis of the body (y-axis).

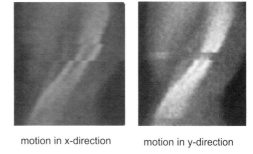

motion in x-direction   motion in y-direction

Figure 7-13 Effects of patient motion on sinograms of a slice of the heart.

Signal vs. Noise

The **signal** is that part of the information that produces the actual image; **noise** is extraneous data and may have no direct relation to the actual image. Noise reduces the quality of the image. Photon scattering, statistical variation, and random electronic fluctuations are among the sources of noise, which can be reduced by improved collimation, longer acquisition times, and better design of the circuitry.

A type of noise peculiar to the reconstruction process is the **star artifact**, so named because a star composed of the backprojection "rays" surrounds each object (Fig. 7-18A). Increasing the number of projection views can reduce this artifact and improve definition (Fig. 7-18B, C). The star artifact can also be reduced by a mathematical technique called filtering, as described below.

Filtering

Filtering is a mathematical technique applied during reconstruction to improve the appearance

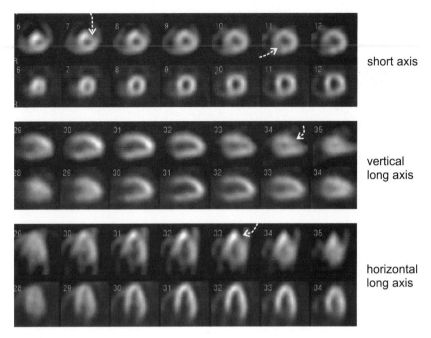

short axis

vertical long axis

horizontal long axis

Figure 7-14 First, third, and fifth rows: artifacts created by patient motion (arrows). Second, fourth, and sixth rows: corrected images.

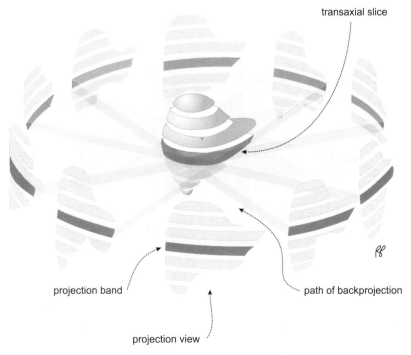

transaxial slice

projection band

path of backprojection

projection view

Figure 7-15 Projection views of a liver are backprojected to create transaxial slices.

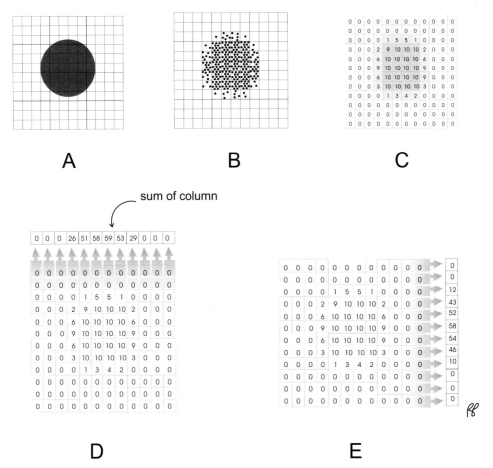

Figure 7-16 Acquistion of projection views (as numerical arrays) of a disk.

of the image. In particular, for our purposes, filters are used to reduce the effects of the star artifact and to remove noise due to photon scattering and statistical variations in counts, as just discussed.

When image data is represented in familiar terms, such as in counts per pixel, this data is said to exist in the **spatial domain**. Filtering can be performed on this data as it is. Alternatively, the data can be represented as a series of sine waves and the filtering performed on these. In the latter case, the data is said to be transformed into the **frequency domain**. Before we discuss this transformation of data from the spatial domain to the frequency domain, we will apply some simple filters to data represented in the spatial domain

in order to reduce the star artifact and noise in images.

Filtering in the Spatial Domain
Spatial Filtering to Reduce Noise: Nine-Point Smoothing

Figure 7-19 demonstrates the effects of noise from scatter and statistical variation on the images of two rectangles. In Figure 7-19A, a grid is placed over the two rectangles being imaged. Figure 7-19B is a representation of an ideal image with uniformly distributed counts. Figure 7-15C is a representation of the effects of scatter and statistical variation of counts in the image. The counts are greater over the rectangles than the background, but scatter and statistical variation

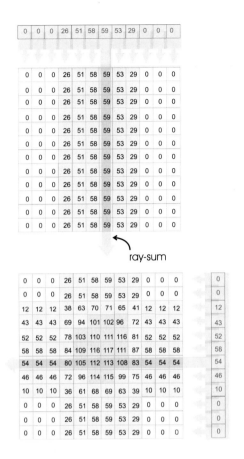

ray-sum

Figure 7-17 Projection views of a disk are backprojected.

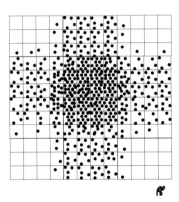

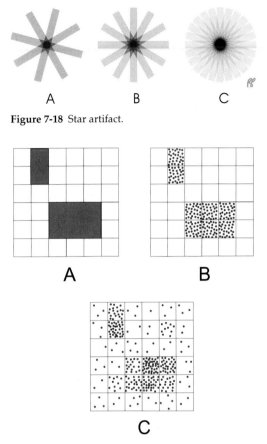

Figure 7-18 Star artifact.

Figure 7-19 Scatter and statistical variation in counts.

contribute to inhomogeneity in uptake in the rectangles and background.

Smoothing partially redistributes the counts from the pixels with the highest counts to its immediate neighbors. In this way the more extreme irregularities in pixel counts are "blunted." The nine-point smoothing technique is demonstrated in Figure 7-20. The counts from the central pixel and eight immediately adjacent pixels are averaged (Fig. 7-20A). The count in the central pixel is replaced by this average. This process is repeated pixel by pixel (Fig. 7-20B, C).

In one variation of this technique, the central pixel is given a different weight than its eight immediate neighbors. This smoothing filter is applied to pixel (2,2) in Figure 7-21A. The nine pixel elements centered on pixel (2,2) are multiplied pixel by pixel by the corresponding elements in the so-called filter kernel (central weight of 10) as shown in Figure 7-21B. The sum of all nine values in the resulting matrix (in this case 339) is divided by the sum of the

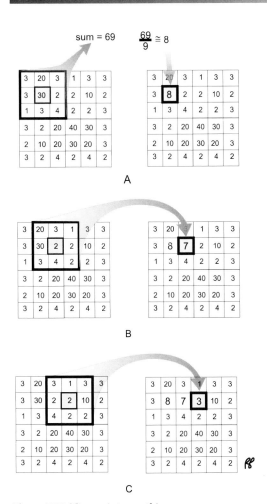

Figure 7-20 Nine-point smoothing.

nine elements of the filter kernel, 18. The resulting value, 19, is the filtered value for pixel (2,2) (Fig. 7-21C). In this way, the original value of pixel (2,2), 30, is reduced by factoring in the lower values of its nearest neighbor pixels. This process is applied to each pixel of matrix.

The result of applying this process to each pixel in the matrix can be seen in Figure 7-22 (upper panel). If a kernel with a less heavily weighted central value of 2 is applied to the matrix, the final image is "more smoothed," that is, the edges of the rectangle are less distinct (lower panel).

Spatial Filtering to Reduce the Star Artifact
A somewhat different effect can be achieved by a technique similar to that just described,

except that the kernel is given negative values for the peripheral pixels and a positive value in the center. This filter tends to enhance the edges and reduce the intensity of the star artifact. A simple version of this filter can be applied to the previous example depicted in Figure 7-17. This kernel consists of a central value, +2, surrounded by values of −1 (Fig. 7-23). This kernel is sequentially applied to each pixel of the array. In the resulting array, the outer values are zero or negative. In a similar fashion, the kernel is applied to the second array of the example in Figure 7-17. When these filtered arrays are backprojected, their peripheral negative values cancel counts in a manner that removes the portion of the rays adjacent to the image of the disk (Fig. 7-24). The relative depression of counts surrounding the backprojected disk helps to separate it from the background.

Figure 7-25 is a graphic representation of this process. The top panels demonstrate the process of backprojecting rectangles to create a disk. Each swipe of the paint roller represents a ray. In the upper right image, the combined rays create a disk with indistinct edges. The bottom images demonstrate the effects of a simple edge enhancement filter in which negative values are used to border each rectangle prior to backprojection (represented by the small white squares on either side of each rectangle). These negative values cancel contributions from adjacent ray-sums and the circle's edge is seen more clearly (bottom right). The filter kernel used here is similar to the kernel in the prior example.

Filtering in the Frequency Domain
Filtering in the spatial domain proves to be computationally burdensome. In general it is easier to perform filtering in the frequency domain, once the data has been transformed. The following is meant to clarify this process.

Until this point we have discussed data only in the most familiar terms, usually as counts per unit time or counts per pixel—in other words, in the **time domain** or in the **spatial domain**.

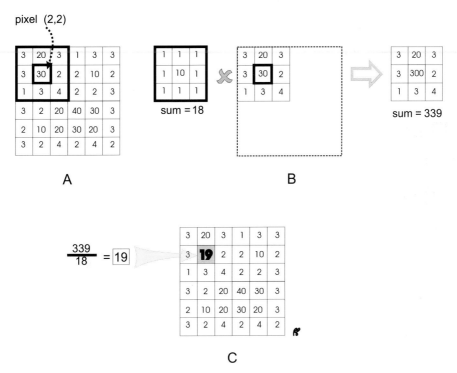

Figure 7-21 Weighted nine-point smoothing.

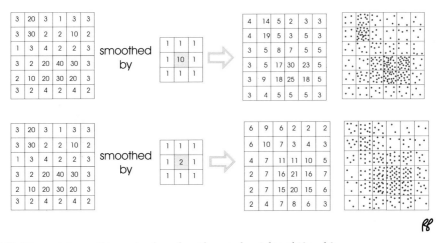

Figure 7-22 Nine-point smoothing using kernels with central weights of 10 and 2.

We may look at the data principally in terms of time, for example, the 24-hour uptake in the thyroid gland, or we may view an image of the spatial distribution of the activity in the thyroid. Although they serve different purposes, these two domains are not entirely independent.

In fact, they only represent different views of the underlying data.

Now we propose to extend this concept of domains beyond the familiar ones of time and space to another—the **frequency domain**. In this rather unfamiliar domain, the distribution

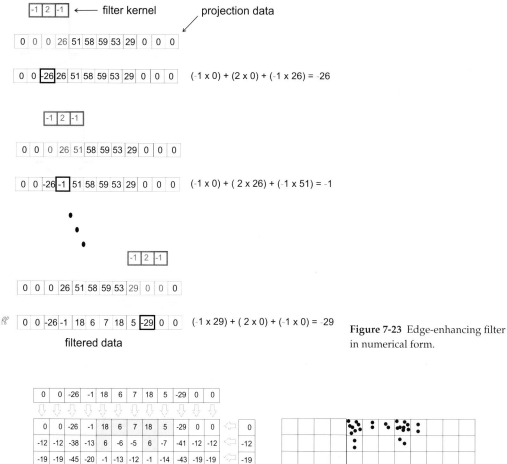

Figure 7-23 Edge-enhancing filter in numerical form.

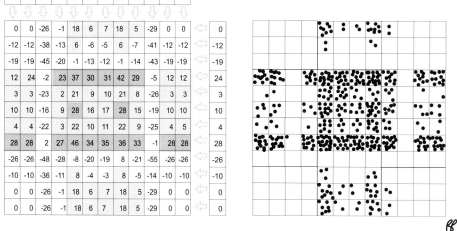

Figure 7-24 Backprojection following application of an edge-enhancing filter.

of counts across the image is expressed by a spectrum of spatial frequencies. The frequencies themselves are given in **cycles per centimeter** or **cycles per pixel**. This transformation of data facilitates the computations necessary for filtering.

The first step is the representation of an object as a sum of sinusoidal waves rather than the arrangement of small dots that we usually refer to as an image in the spatial domain. Just as any sound pattern in air can be represented by a combination of sine and cosine waves given

Figure 7-25 Graphic representation of an edge-enhancing filter.

in cycles per second, any image pattern can be represented by a combination of sine and cosine waves in cycles per pixel. This use of sinusoidal waves to represent a simple object is illustrated in Figure 7-26. The original rectangles (seen in the top row of column C) can be plotted as a square wave (column A). The three-dimensional view of this square wave is drawn in column B. Column C can be thought of as a bird's-eye view from the top of the three-dimensional square wave. The subsequent rows depict the process of the sequential addition of sine waves to approximate the square waves. The first sine wave (seen in column A of the second row) is a very rough approximation of the square wave and when viewed from above poorly represents the original

rectangles. A second sine wave (of higher frequency) is added to the first sine wave (shown in the third row). Each subsequent addition of a higher frequency sine wave serves to sharpen the image of the rectangles in column C.

In contrast to the original dot image, which can be described by the number of dots at each location, the amplitude of the wave at each frequency now describes the new image (Fig. 7-27). This would be the object (in our example, the two rectangles) described in the frequency domain, and a plot of amplitudes of the wave at each frequency is called its **frequency spectrum** (Fig. 7-28). Amplitude, the height of the wave, is expressed in counts; frequency is measured in cycles per pixel.

original data

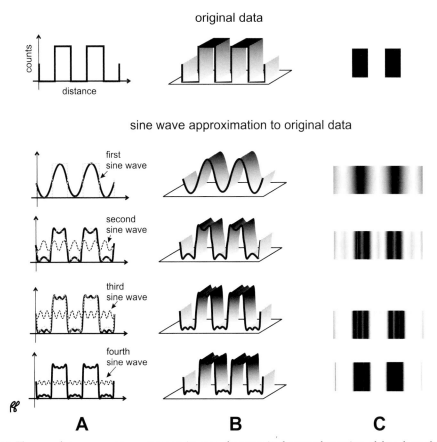

sine wave approximation to original data

A   **B**   **C**

Figure 7-26 The use of sine waves to represent an image: a key step in the transformation of data from the spatial to the frequency domain.

The information obtained by the camera is not changed by this transformation of the collected data from the spatial to the frequency domain; all that is changed is the method of describing the data. More generally, we can say that data can be transformed from one domain into another with neither gain nor loss of the contained information.

While we are interested here in mathematical methods of transforming data, it is worth mentioning that common physical devices, optical lenses for example, can also be used to transform data. The most pertinent data transformation from the spatial to the frequency domain is based on a method developed about 200 years ago by the French mathematician and physicist J.B.J. Fourier and is now referred to as the **Fourier**

transformation. Its importance lies in the computational time it saves when filtering image data.

Nyquist Frequency

The highest fundamental frequency useful for showing that two adjacent points are separate objects is 0.5 cycles/pixel, which can be demonstrated by showing the effect of higher frequencies. In the top panel of Figure 7-29, a frequency of 0.25 cycles/pixel will cause every fourth pixel to turn on. For demonstration purposes, we have assumed here that only the most positive portion of any sine wave can turn on a pixel. For the frequency of 0.5 cycles/pixel, every other pixel will be on. At still higher frequencies (1.0 cycles/pixel, 2.0 cycles/pixel, and so on), every pixel will be on; in this situation one

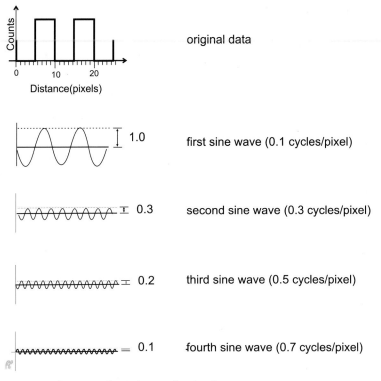

original data

first sine wave (0.1 cycles/pixel)

second sine wave (0.3 cycles/pixel)

third sine wave (0.5 cycles/pixel)

fourth sine wave (0.7 cycles/pixel)

Figure 7-27 Sine waves used to approximate image of rectangles.

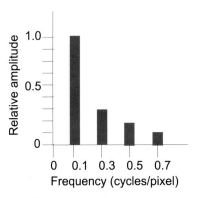

Figure 7-28 Frequency spectrum.

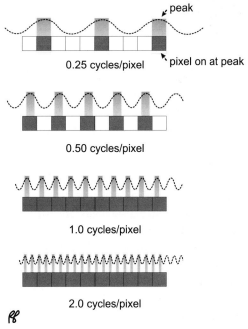

Figure 7-29 The Nyquist frequency of 0.5 cycles/pixel is the smallest discernible frequency for a matrix.

sine wave peak cannot be separated from the next. The frequency of 0.5 cycles/pixel is referred to as the **Nyquist frequency**; it is the highest fundamental frequency useful for imaging.

Although the Nyquist frequency is always 0.5 cycles/pixel, when it is expressed as cycles/cm the numeric value is a function of the pixel size. The smaller the pixel size the

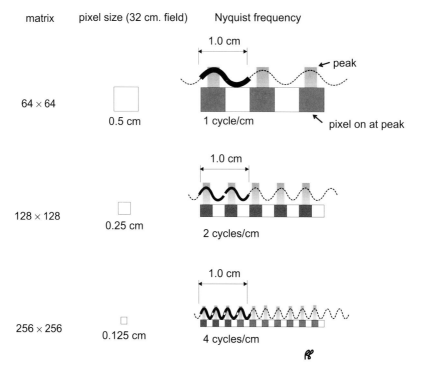

matrix   pixel size (32 cm. field)   Nyquist frequency

64 × 64

0.5 cm   1 cycle/cm

1.0 cm

peak

pixel on at peak

128 × 128

0.25 cm

1.0 cm

2 cycles/cm

256 × 256

0.125 cm

1.0 cm

4 cycles/cm

Figure 7-30 The Nyquist frequency expressed in cycles/cm.

greater the Nyquist frequency in cycles/cm. As shown in Figure 7-30, for a pixel size of 0.5 cm, 0.5 cycles/pixel is 1.0 cycle/cm; for a pixel size of 0.25 cm, it is 2 cycles/cm; and so on.

Signal, Noise, and the Star Artifact in the Frequency Domain

To understand the design of filters it is necessary to look at the frequency distribution of the important components of an image. Figure 7-31A is a distribution of frequencies derived from a hypothetical image; the image itself is not shown. The darker gray bars derive from the signal data that are composed principally of frequencies in the low to middle range; white bars represent statistical noise, which is nearly uniform across the spectrum; the light gray bars derive from the data producing the star artifact and are principally in the low range of frequencies. An ideal filter removes all noise data and retains all signal data. Unfortunately, the frequency ranges of signal and noise overlap. Filters are designed to optimize the signal in the presence of noise.

Frequency Filtering to Reduce the Star Artifact

The **ramp** filter is named for its shape in the frequency domain. This filter was designed to reduce the star artifact resulting from backprojection. Figure 7-31B demonstrates the effect of the ramp filter on the star artifact; notice that it removes more data at the low end of the spectrum.

> ### WINDOW
>
> Another name of the low-pass filters used in nuclear medicine.

Frequency Filtering to Reduce Noise

Filters can be described by the portion of the frequency spectrum that they transmit. The most common examples are the low- and high-pass filters. Low-pass filters reject high-frequency data; high-pass filters reject low-frequency data. The ramp filter just described is an example of a high-pass filter.

In general, the use of a low-pass filter results in an image with indistinct edges and loss of detail

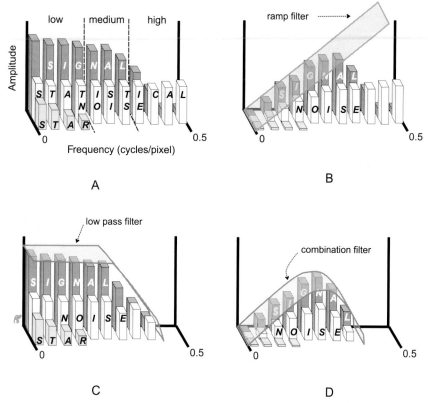

Figure 7-31 Effect of the ramp and low-pass filters on signal data, statistical noise, and the star artifact.

(Fig. 7-32, top panel). High-pass filters accentuate edges and retain finer details but can be difficult to interpret due to their "grainy" appearance caused by high-frequency noise (Fig. 7-32, bottom panel).

Low-Pass Filters

Although the ramp filter is an excellent means of removing the star artifact and also serves to reduce some of the low- and mid-frequency noise, the image still contains a large quantity of high-frequency noise that can interfere with interpretation of the images. High-frequency noise from statistical variation in counts is more of a problem in low-count images, such as SPECT projection images, than it is in a standard high-count planar view. A **low-pass filter** is used to reduce the contribution of high-frequency noise. The effect of a low-pass filter

on our frequency distribution can be seen in Figure 7-31C.

Types of low-pass filters: There are many low-pass filters available to process nuclear medicine data, and all are named after their inventors, for example, Hann (or vonHann), Hamming, Butterworth, Weiner, and Parzen. Each filter has a different shape (albeit some are quite similar) and when applied will modify the image differently.

The effects of some typical low-pass filters used to reduce high-frequency noise are plotted in Figure 7-33. The Parzen filter is an example of a low-pass filter that greatly smoothes data; generally it is not used for SPECT. The Hann and Hamming filters are low-pass filters with some smoothing but a relatively greater acceptance of mid- and high-frequency data than the Parzen filter. The commonly used Butterworth

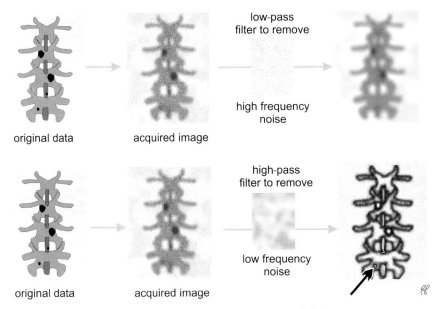

Figure 7-32 Low-pass and high-pass filters. Black arrow points to preserved details.

filter allows the user to adjust the relative degree of high-frequency wave acceptance. The Butterworth and Hann filters are also "flexible" filters in that their shape can be altered by specifying certain parameters. The very light gray broad band in Figure 7-33 roughly delineates the possible range of shapes of the Hann filter; the darker gray band delineates possible shapes of the Butterworth window.

Cutoff frequency and order: The **cutoff frequency**, often referred to as the power of the filter, is the maximum frequency the filter will pass. If the cutoff frequency is greater than the Nyquist frequency, the filter is abruptly terminated at the Nyquist frequency of 0.5 cycles/pixel. A Hann filter is depicted with different cutoff frequencies in the upper panel of Figure 7-34. An additional parameter, the **order**, can be specified for Butterworth filters. The order controls the slope of the curve (lower panel of Fig 7-34).

Sequence for Applying Filters

Filters can be applied to the data prior to backprojection (**prefiltering**), during backprojection, or to the transaxial slices following backprojection. Often the low-pass filters (Butterworth, Hann, Hamming, Parzen, and others) are applied as prefilters to remove high-frequency noise. The ramp filter is then applied during backprojection to remove the star artifact and incidentally remove other low-frequency data. The prefilter and ramp filter can be applied in a single step (see Fig. 7-31D). This **combination filter** is either referred to by the filter's name, for example "Parzen," or by a description such as "Ramp-Parzen filter."

Filter Selection

The selection of the optimal filter depends on both the characteristics of the data and on the user's personal preference. In general, a high-pass filter is better suited for higher count data, whereas a low-pass or smoothing filter is better for data containing a small number of counts. With respect to preference, some users like smoother, less-detailed images; others are willing to tolerate the "grainy" appearance following the use of a high-pass filter in order to retain the finer details of the signal data.

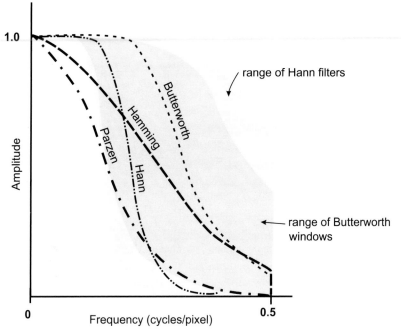

Figure 7-33 Characteristics of commonly used low-pass filters.

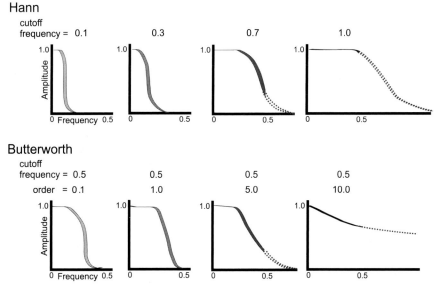

Figure 7-34 The Butterworth and Hann windows (or prefilters) can be modified to match the characteristics of the data set.

Attenuation Correction

Photons originating from inside the body are more likely to be absorbed or scattered by the surrounding tissue than photons originating near the surface. The attenuation coefficient for ^{99m}Tc in tissue (see Chapter 2) is 0.15/cm. This means that every centimeter of tissue between the source

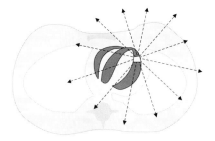
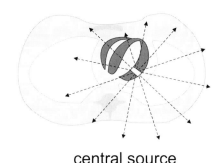

superficial source   central source

Figure 7-35 Attenuation of photons.

and camera will absorb or scatter approximately 15% of entering photons. Attenuation of photons will reduce counts from the middle of the body. In Figure 7-35 the photons from the deeper portions of the myocardium are less likely to exit the surface of the body than photons from the more superficial portions.

Attenuation correction in the past was routinely performed by calculation techniques which assumed uniform tissue density across the body and therefore uniform attenuation. Transmission imaging for measuring attenuation, which is rarely uniform across the body, is gradually replacing calculation techniques, especially in the chest where soft-tissue, lung, and bone densities coexist in the same transaxial slices.

Calculated Attenuation Correction

Attenuation correction can be performed by applying a correction factor that takes into account source depth and the tissue attenuation coefficient. The attenuation correction factor is assumed to be constant throughout the cross-section of the body. This assumption is more accurate in the head and abdomen where tissue is more uniform; it is less accurate in the chest with its mix of aerated lung and mediastinal and chest wall tissue. The correction factor for the head is approximately $0.13/cm$ and for the abdomen is $0.12/cm$; these values are less than the theoretical attenuation coefficient of $0.15/cm$ because

scattered photons within the tissue "falsely" increase the counts in the center of the body.

There are several algorithms available for calculated attenuation correction. A simplified application of one such mathematical technique, the Chang algorithm, is the most commonly used method. Figure 7-36 highlights the major points of the method. (Note that for illustration purposes an image of the thorax is used; in general, however, calculated attenuation correction is reserved for the denser organs, such as the brain and liver). An approximation to the outline of the thorax is drawn by computer or manually (Fig. 7-36A). Within the outline, a correction matrix is constructed; this is shown symbolically by the shading (Fig. 7-36B) in which the darker area indicates greater correction. As seen in Figure 7-36C, the greater correction increases the number counts in the deeper portions of the thorax.

Transmission Attenuation Correction

In this technique correction matrices are reconstructed from **transmission images**. Transmission images are tomographic projection images obtained from a radionuclide source positioned outside the patient on the side directly opposite the camera (Fig. 7-37). Attenuation of photons through the body from this external source is dependent on tissue thickness. As the camera head and source are rotated around the patient, the attenuation of photons through the body

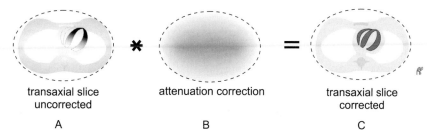

transaxial slice uncorrected	attenuation correction	transaxial slice corrected
A	B	C

Figure 7-36 Attenuation correction.

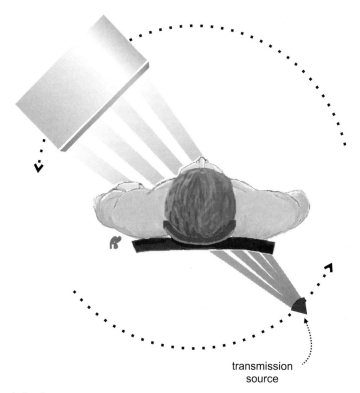

transmission source

Figure 7-37 Transmission image.

varies with the thickness of the tissue between the source and the camera head. A 360° set of images are obtained and the transaxial slices, referred to as transmission slices, are constructed from this set of images in the standard method. From this set of images it is possible to estimate the amount of attenuation, which in turn is used to correct patient images. Transmission imaging for attenuation correction is also used in PET and PET-CT and will be covered in more depth in Chapter 8.

Iterative Reconstruction

A relatively elegant technique called **iterative reconstruction** is steadily replacing filtered backprojection. Images reconstructed with this technique exhibit significantly less star artifact (see Figure 7-18) than those created using filtered backprojection.

In iterative reconstruction, the computer starts with an initial "guess"-estimate of the data to produce a set of transaxial slices. These slices are then used to create a second set of projection views which are compared to the original projection views as acquired from the patient. The transaxial slices from the computer's estimate are then modified using the difference between, or ratio of, the two sets of projection views. A new set of transaxial slices reconstructed from this modified, or second, estimate are then used to create a set of projection views which are compared to the original projection views. Again these projection views are compared to the original projection views. If the process proceeds efficiently, each iteration generates a new set of projection views that more closely approximate the original projection views. The process is complete when the difference between the projection views of the estimated data and the original data is below a pre-determined threshold.

Figure 7-38 is a very simplified representation of this process using a single transaxial slice shaped like a squared-off horseshoe, to represent a transaxial slice of a heart, and a simple square object to represent the computer's initial estimate of this transaxial slice. For purposes of simplification the "projection" views in Figure 7-38 are not true projection views but merely surface imprints of the transaxial slices. Figure 7-38A illustrates the creation of the original and estimated data projection views. Figure 7-38B illustrates the comparison of the projection views for the first estimate with those of the original data and the subsequent creation of the correction projection views from their ratio. A transverse correction slice is then constructed by "backprojecting" these projection views. The estimate of the first slice is then multiplied by this correction slice to create the second estimate. Figure 7-38C demonstrates the same process as illustrated in Figure 7-38B using the transaxial slice from the second estimate. Lastly, Figure 7-38D summarizes the results for five iterations of this idealized model; at the fifth iteration the estimated transaxial slice equals the original data and the process is complete.

In practice most iterative reconstructions are terminated at a pre-determined number of iterations, that is, when the radiologist is satisfied with the overall image quality, instead of allowing them to progress until the difference between the estimated and projection views reaches a set value. In general the image resolution improves with increasing number of iterations as demonstrated in Figure 7-39 using a single coronal slice of a lumbar spine from a SPECT bone scan. However, beyond a certain reasonable number of iterations, further improvements in resolution can only be accomplished at the cost of increased image noise.

Another advantage of iterative reconstruction compared to filtered backprojection is that corrections for attenuation, scatter, and even collimator and detector spatial resolution can be incorporated directly. The use of iterative reconstruction techniques was limited in the past because of lengthy computation times; faster computers and "short-cut" mathematical techniques such as OSEM have mitigated this problem.

OSEM

One widely used iterative reconstruction technique is the **maximum likelihood expectation maximization (MLEM)** algorithm. The computation time is very lengthy when all of the projection views are used in each iteration to create the correction slices. To shorten the processing time a smaller group, or **subset**, of matched projection views from the estimated and original datasets are used for each iteration to create the correction slices. For each iteration a different group or subset is used. Each subset consists of projection views sampled evenly over the entire arc of the acquisition, 180° or 360°. For example, the first subset might contain every twelfth projection view beginning with the first projection view. The second subset might contain every twelfth projection view beginning with the sixth projection view, and so on. The process of using subsets for expectation maximization is called **ordered subsets expectation maximization (OSEM)**.

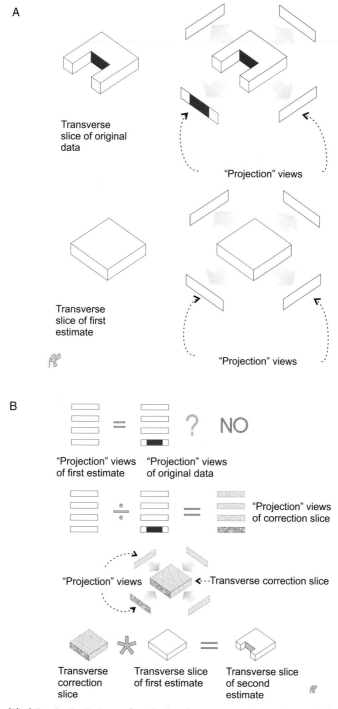

Figure 7-38 (A) Simplified "projection" views of original and computer estimated transaxial slices. (B) First iteration. (C) Second iteration. (D) Progression of five iterations.

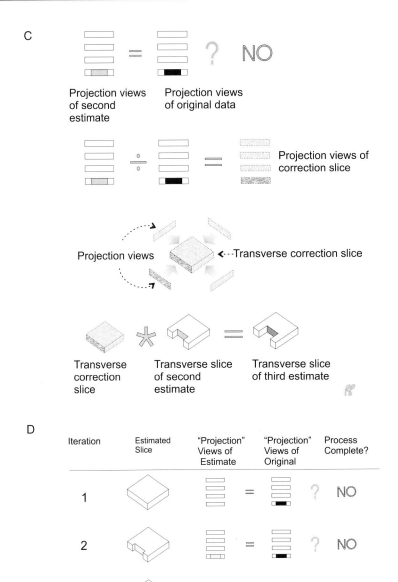

Figure 7-38 Continued.

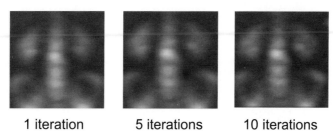

1 iteration   5 iterations   10 iterations

Figure 7-39 Resolution improves with increasing iterations.

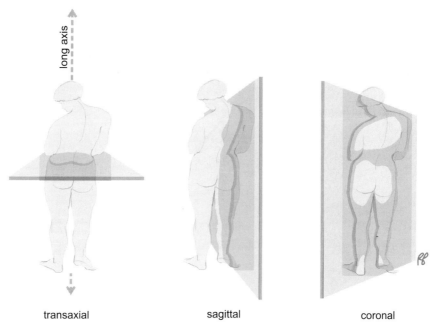

transaxial   sagittal   coronal

Figure 7-40 Transaxial, sagittal, and coronal images.

Post-reconstruction Image Processing

Creation of Sagittal, Coronal, and Oblique Slices

Once the transverse images have been constructed **sagittal**, **coronal**, and if necessary oblique views can be generated. By convention, the transaxial slices are oriented perpendicular to the long axis of the body. Sagittal and coronal slices are oriented parallel to the long axis of the body and at right angles to each other (Fig. 7-40). In a similar way, the axes of the heart are defined relative to the heart's long axis, understood to be the line running from base to apex. The **horizontal long axis view** and **vertical long axis view** are oriented parallel to the long axis of the heart. The **short axis view** is oriented perpendicular to the long axis of the heart (Fig. 7-41).

Advanced Display Techniques

Three-dimensional displays of SPECT data can aid in interpretation and presentation of data. Standard methods of display, including surface rendering (visualization of the surface of the data), volume rendering (a more transparent view of the data set), and projection views

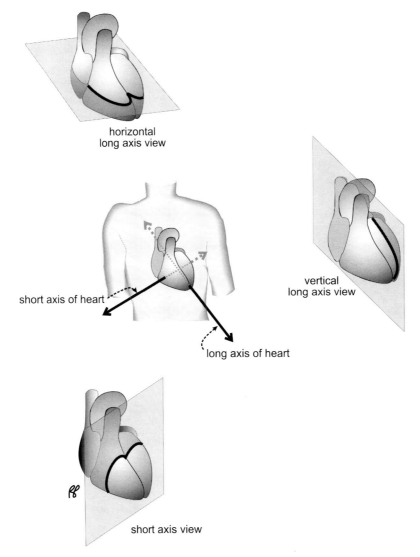

horizontal
long axis view

short axis of heart

long axis of heart

vertical
long axis view

short axis view

Figure 7-41 Oblique views of the heart.

(greatly enhanced cine views), are available on different systems. An in-depth discussion of these methods is beyond the scope of this text.

Test Yourself

1 Match the following phrases to the terms listed below:
  (a) Individual images acquired at each stop of the SPECT camera as it rotates around the patient

  (b) "Movie" of all of the projection views
  (c) Reconstructed horizontal slices of the patient data
  (d) A stack of slices from each acquired projection view, each slice taken at the same level of the body

Terms:
  (i) cine view
  (ii) projection views
  (iii) sinogram
  (iv) transverse slices.

2 Which of the following statements about filtering used in reconstruction is correct:
- **(a)** Filtering is used to reduce noise and preserve signal data in the image.
- **(b)** Filtering cannot be applied to the data prior to backprojection.
- **(c)** Filtering is used to reduce the star artifact created during backprojection.
- **(d)** All of the above.

3 True or false: the frequency domain represents an image as a sum of sine waves.

4 Match the following statements with either "low pass filter" or "high pass filter":
- **(a)** Used to remove high frequency noise such as noise from statistical variation in counts
- **(b)** Used to remove low frequency noise such as the star artifact
- **(c)** Can create "grainy" images, but preserves details
- **(d)** Creates "smoother" images, but image detail may be lost.

5 Which of the following statements about filters is correct?
- **(a)** The Nyquist frequency is 5.0 cycles/pixel.
- **(b)** The cutoff frequency of a filter is the minimum frequency the filter will pass.
- **(c)** Nyquist filters have an additional parameter called the order of the filter which controls the slope of the curve.
- **(d)** All of the above.
- **(e)** None of the above.

6 Which of the following statements are true about iterative reconstruction
- **(a)** Images reconstructed with iterative reconstruction have less star artifact than those reconstructed with filtered back projection
- **(b)** Attenuation correction can be applied during iterative reconstruction
- **(c)** Due to long computational times "short-cut" techniques such as ordered subsets expectation maximization are often used in iterative reconstruction
- **(d)** Successive iterations are performed until the projection views of the estimate closely approximate the original data or until a pre-defined number of iterations have been completed.
- **(e)** All of the above.
- **(f)** None of the above.

7 True or false: Measured tissue attenuation using transmission sources is generally more accurate than calculation techniques for attenuation correction which assume uniform tissue density across the body.

8 Positron emission tomography (PET)

Positron emission tomography (PET) cameras are designed to detect the paired 511-keV photons generated from the annihilation event of a positron and electron. Following emission, any positron travels only a short distance before colliding with electrons in surrounding matter. As discussed in Chapter 2, the paired 511-keV annihilation photons travel in opposite directions (180° apart) along a line (Fig. 8-1).

Following the acquisition of the images of positron emissions, the data are reconstructed in a manner similar to that used for SPECT, with the exception that attenuation correction is always performed and usually with the transmission scan method (see Chapter 7) or with

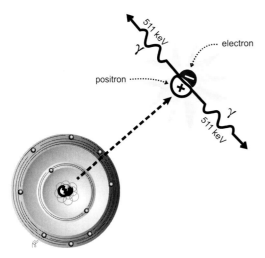

Figure 8-1 Annihilation reaction.

the incorporation of CT data. PET has a number of advantages compared to SPECT. Most important are its greater sensitivity and resolution and the existence of positron emitting isotopes for elements of low atomic number elements for which no suitable gamma emitters are available. The principal disadvantage of PET is the added cost of the equipment and the short half-life of some of the most useful positron emitters.

Advantages of PET Imaging

Sensitivity

As pointed out earlier, a collimator reduces a camera's sensitivity because the collimator's septa cover part of the camera crystal's face. A collimator is not required for PET since PET cameras use the detection of the simultaneous and oppositely directed 511-keV photons of positron annihilation to locate the direction from which the photons originated. This is referred to as "**annihilation coincidence detection**." For SPECT, the origins of the single photons can only be located if the camera is equipped with a collimator to absorb any photons not traveling perpendicular to the face of the crystal. Therefore, a PET camera, needing no collimator, is inherently more sensitive (by at least a factor of 100) compared to a SPECT camera and has a higher count rate for similar quantities of radioactivity.

Resolution

Coincidence Detection

If two detectors located on opposite sides of the annihilation reaction register coincident photon impacts, the annihilation reaction itself occurred along an imaginary line, called a **line of response (LOR)**, drawn between these detectors (Fig. 8-2). A **coincidence circuit** registers these as simultaneous events.

Events that arise from a single positron annihilation that follow the emission of a positron are referred to as **true coincidence events**. The impact of an unpaired photon, called a **singles event**, is rejected. A singles event is registered when an unpaired photon from a non-annihilation gamma ray impacts a detector (labeled A in Fig. 8-3). A singles event is also registered when only one of a pair of annihilation photons impacts a detector; the other photon can leave the plane of detection (B in Fig. 8-3) or it can be absorbed or scattered by the surrounding medium (C in Fig. 8-3).

Unfortunately, by chance, photons generated simultaneously from separate sites in the body may reach the crystal at the same time. These separate events are incorrectly perceived as though resulting from annihilation of a single positron

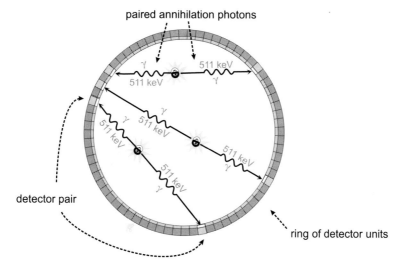

Figure 8-2 Examples of coincident events.

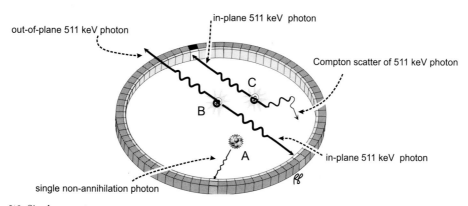

Figure 8-3 Singles events.

emission occurring along a line between the two detectors. These photons can arise from annihilation events (Fig. 8-4A), when one of each pair of two annihilation events registers as a simultaneous event (their respective partners are out-of-plane events) or from simultaneous detection of non-annihilation photons (Fig. 8-4B). Other mistakes in identification may also occur. For example, when scattering of one of the 511-keV photons alters it's path, the location of the annihilation event may incorrectly be presumed to be on a line connecting the detectors (Fig. 8-4C). These are referred to as **random events** (a sort of

false positive happening). The probability of random events increases significantly with increasing radioactivity within the field of view of the scanner and thus are of most concern at high count rates. Scatter and random events are undesirable because they contribute to an increase in image background counts and consequently cause a reduction in image contrast.

Time of Flight

To improve resolution, some systems also measure **time of flight** under the assumption that the location of the annihilation can be localized along

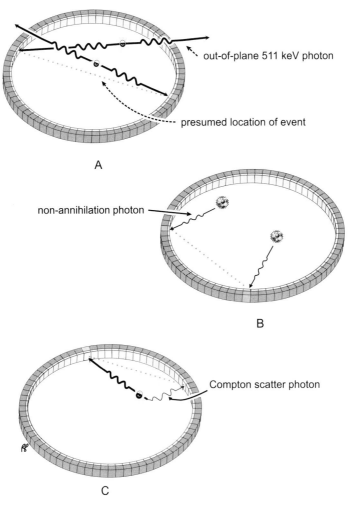

Figure 8-4 Random events.

the line of flight of the coincident photons by measuring the time of arrival of each of the photons at the opposing crystals. Unless the event occurs in the exact center of the detection ring, one of the photons will arrive before the other. The time difference will be proportional to the difference in distances traveled by the two photons and can be used to calculate the position of the event along the line connecting the detectors (Fig 8-5). Unfortunately, due to electronic timing limitations, the calculated position is not accurate and the resulting reduced spatial resolution of this technique has minimized its use, although further advances in the technology may allow its use in the future.

Radiopharmaceuticals

One of the greatest advantages of PET imaging is the large number of low atomic number elements for which positron emitters exist (Table 8-1). This permits incorporation of positron emitters into many biologically active compounds, including isotopic forms of oxygen, carbon, nitrogen, and fluorine. Very specific physiologic properties of an organ can therefore be

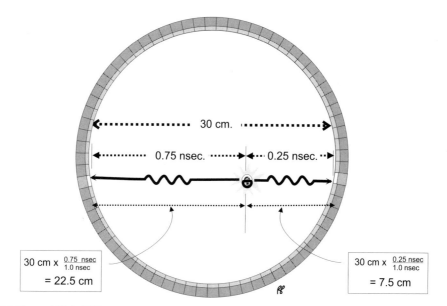

Figure 8-5 Time-of-flight PET systems.

Table 8-1 Common Positron-Emitting Nuclides Update

Nuclide	Half-Life (min)	Positron Yield (%)	Maximum Energy (MeV)	Method of Production
^{11}C	20.4	99.0	0.960	Cyclotron
^{13}N	9.96	100.0	1.190	Cyclotron
^{18}F	110.00	97.0	0.635	Cyclotron
^{15}O	2.04	99.9	1.720	Cyclotron
^{82}Rb	1.27	96.0	3.350	Generator
^{62}Cu	09.8	98.0	2.930	Generator
^{68}Ga	68.1	90.0	1.900	Generator

imaged; for example, the oxygen consumption and the glucose metabolism of the brain can be independently imaged using $[^{15}O]_2$ and ^{18}F-fluorodeoxyglucose. Other radiopharmaceuticals that can be used in PET are listed in Table 8-2.

PET Camera Components

A common PET detector consists of rings of crystals (Fig. 8-6). The rings may or may not be separated by septa. The individual detector contains one or more large segmented crystals or a collection of small crystals. A standard **detector unit** or **block** consists of a small crystal or small portion of a larger crystal watched by four photomultiplier tubes.

Crystals

The basic crystal function of converting gamma photon energy into light photon energy is discussed in Chapter 5. Thallium-doped sodium iodide (NaI(Tl)) crystals were originally used for PET systems; however, NaI has a relatively low density and is less effective at stopping the high-energy 511-keV photons. To compensate for the lower density, thicker crystals were employed for 511-keV imaging than those used for detecting the lower-energy single photon emissions (3.8 cm as opposed to 0.6 to 1.2 cm).

Crystals with higher densities and higher atomic numbers (Z), such as bismuth germinate oxide (BGO), leutetium orthosilicate (LSO), and gadolinium orthosilicate (GSO), are now commonly used for 511-keV imaging due to their greater sensitivity for detecting photons than less dense crystals. This is because the likelihood of a photoelectric interaction increases rapidly with increasing Z; the increase is proportional to Z^3. In addition, in a denser material, a Compton interaction is also more likely to occur than in a less dense material because electrons are more numerous in a given volume of a dense material. Compton interactions, however, are much less common than photoelectric interactions in materials with high atomic numbers (see Fig. 2-1).

LSO and GSO have the advantage of a shorter **decay time** than BGO or NaI(Tl) for PET imaging. Decay time is the time required for the radiation-excited molecules of the crystal lattice to return to the ground (unexcited) state with the emission of light photons (see Fig. 5-2 and associated text). During this time, a second gamma photon entering the crystal cannot be detected, so the longer the decay time the fewer the number of photons that can be detected and the lower the sensitivity of the crystal. A shorter decay time is desirable

Table 8-2 Some of the Available PET Radiopharmaceuticals Update

Radiopharmaceutical	Physiologic Imaging Application
$[^{15}O]_2$	Cerebral oxygen metabolism and extraction
$H_2[^{15}O]$	Cerebral and myocardial blood flow
$C[^{15}O]$	Cerebral and myocardial blood volume
$[^{11}C]$-N-methylspiperone	Cerebral dopamine receptor binding
$[^{11}C]$-methionine	Tumor localization
$[^{11}C]$-choline	Tumor localization
$[^{18}F]$-fluorodeoxyglucose	Cerebral and myocardial glucose metabolism and tumor localization
$[^{13}N]H_3$	Myocardial blood flow
$[^{11}C]$-acetate	Myocardial metabolism
$[^{82}Rb]^+$	Myocardial blood flow

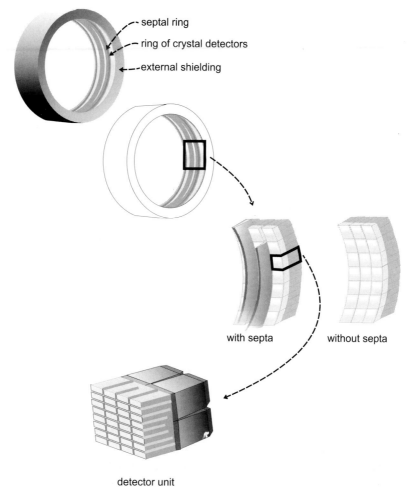

septal ring

ring of crystal detectors

external shielding

with septa   without septa

detector unit

Figure 8-6 PET camera.

for rapid imaging (see the section on dynamic images in Chapter 6).

A higher **light yield** is desirable because the greater the crystal light-photon output per keV of absorbed gamma energy the greater the energy resolution and better spatial resolution from better crystal identification. Improved energy resolution improves the ability to distinguish the lower energy scatter photons from the high-energy annihilation photons. The larger number of light photons makes it easier to identify which crystal has been struck by the annihilation photon. As outlined in Table 8-3, the choice of crystal material involves a compromise among

the factors of cost, density, decay time, and light yield.

Photomultiplier Tubes

Basic photomultiplier tube (PMT) function is discussed in Chapter 5, Figure 5-5. Unlike single photon imaging systems, where many PMT tubes are attached to a single large crystal, positron cameras are designed with many crystal subdivisions watched by a few PMT tubes. The slits between crystal subdivisions channel the light photons toward the PMTs. Localization of the site of impact is achieved by measuring the light

Table 8-3 Properties of Crystals Used for PET Imaging Update

Crystal	Density (g/cm^3)	Decay Time (ns)	Light Yield Relative to NaI (%)
NaI(Tl)[a] (sodium iodide)	3.67	230.0	100
BGO (bismuth germanate oxide)	7.13	300.0	014
LSO (leutetium orthosilicate)	7.40	040.0	075
GSO (gadolinium orthosilicate)	6.71	060.0	041
CsF (cesium fluoride)	4.64	003.0	008
BaF$_2$ (barium fluoride)	4.89	000.8	005

Source: Early, PJ and Sodee, DB, *Principles and Practice of Nuclear Medicine*, CV Mosby, St. Louis, 1995, p. 319; and Ficke, DC, Hood, JT, and TerPogossian, MM, A Spheroid Positron Emission Tomograph for Brain Imaging: A Feasibility Study, *Journal of Nuclear Medicine*, 1996, 37: 1219–1225.

Note:
[a] 3.8 cm thick (compared to 0.6 to 1.2 cm thick for single photon emission cameras).

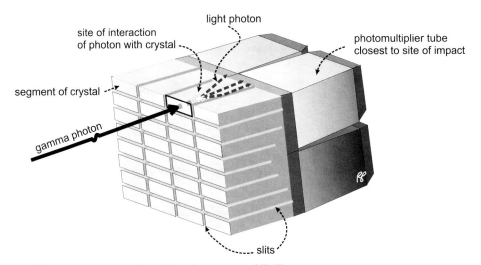

Figure 8-7 Slits between crystals direct light photons toward PMTs.

detected in each PMT; the closer the PMT is to the site of impact the stronger the signal generated by the PMT (Fig. 8-7).

Pulse Height Analyzers, Timing Discriminators, and Coincidence Circuits

The signals from the PMT tubes are amplified by the pre-amplifiers and amplifiers (see Fig. 5-5 and associated text). The system electronics must then determine which signals came from paired 511-keV coincident photons arising from an annihilation event occurring along a line of response between a pair of opposing detectors. This is accomplished primarily by measuring the size of each signal, which is proportional to the energy of the photon reaching the crystal, and by recording the time of detection of the

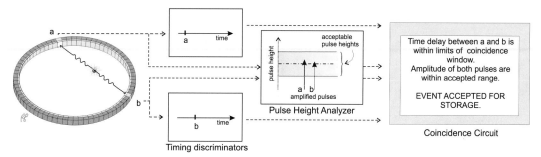

Figure 8-8 A coincident event is accepted after processing by the pulse height analyzers, timing discriminators, and coincidence circuits.

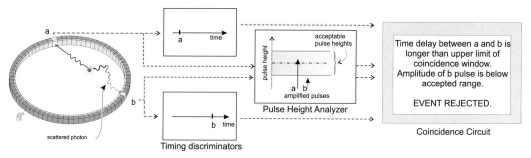

Figure 8-9 One of a pair of annihilation photons is scattered and their data is rejected.

signal. The pulse-height analyzer (Fig. 5-5) determines whether the signal is the correct amplitude (pulse height) to have come from a 511-keV photon interaction within the crystal. The **timing discriminator** records the time the signal was generated. The **coincidence circuit** then examines signals of adequate amplitude coming from opposing detectors and determines if the timing of the signals occurred within the **coincidence time window**. Typically, this coincidence timing window is set between 5–15 ns depending on the decay time of the selected crystal material. As illustrated in Figure 8-8, two 511-keV coincident photons coming from an annihilation event along the line of response between paired detectors will yield adequately sized pulses and a timing delay that is less than the upper limit of the coincidence window. A coincident event between the detector pairs will be recorded by the computer. In contrast, paired annihilation photons in which one or both of the photons have been subjected to scatter or absorption by surrounding tissue will usually be discarded by the coincident circuit, as seen in Figure 8-9. This is because scattered photons have lower photon energies and because in the process of scattering the photon is often delayed on it's path to the crystal. Random events are more difficult to separate from coincident events since they occur as the result of two 511-keV (or other high energy) photons arising from events occurring away from the line of response but reaching the opposing detectors within the coincidence time window (Fig. 8-10) . Some systems estimate random event rates arithmetically using a product of the coincidence time window and measured singles rates. Other approaches are available but are beyond the scope of this text.

Septa

Septal rings can be used to improve resolution by reducing the amount of scatter from photons originating outside the plane of one ring

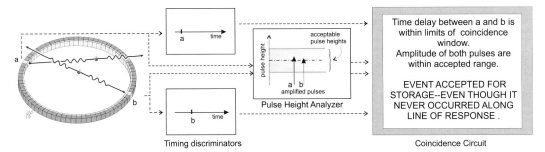

Timing discriminators · Pulse Height Analyzer · Coincidence Circuit

Figure 8-10 Random events can "pass" as true coincidence events.

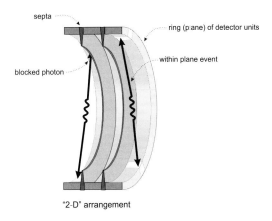

"2-D" arrangement

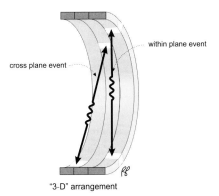

"3-D" arrangement

Figure 8-11 Two-dimensional and three-dimensional PET imaging.

of crystals. The sensitivity of the scanner is reduced, however, because a significant fraction of true coincidence events are rejected. Removal of the septa will increase sensitivity and decrease resolution. Scans obtained with the septa in place are called **two-dimensional scans**; without septa the scans are called **three-dimensional scans**. In the "**2-D**" illustration in Figure 8-11, the septa block out-of-plane photons, allowing only within-plane coincidence events to be recorded. The "**3-D**" configuration permits coincident registration of cross-plane events, those in which the two 511-keV photons are detected in different rings. Septa reduce the number of random coincidence events.

Factors Affecting Resolution in PET Imaging

Positron Range in Tissue

Positrons travel a short distance in tissue before undergoing annihilation with an electron. Therefore, the camera detects photons originating from an annihilation event at a distance from the true source of the beta particle emission (Fig. 8-12A). For lower energy beta emitters (such as ^{18}F), this range is fairly small (1.2 mm in water); for higher energy beta emitters (such as ^{82}Rb), the distance traveled prior to detection can be quite large (12.4 mm in water) [1]. The minimum possible resolving power of any system for a positron nuclide is therefore limited to the average range of the positrons in the tissue.

Photon Emissions Occurring at Other than 180°

Another factor causing degradation in resolution is the fact that 511-keV annihilation photons do not always travel in paths separated by exactly

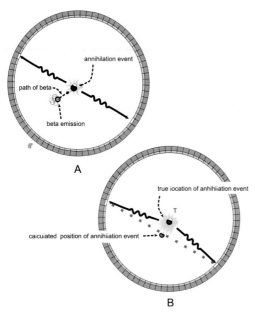

Figure 8-12 (A and B) Factors limiting resolution in PET imaging.

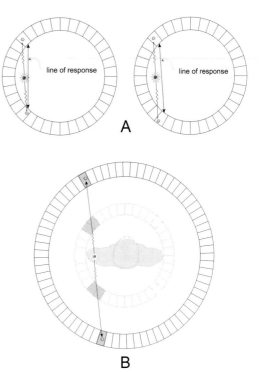

Figure 8-13 (A) Parallax error affects resolution near periphery of field. (B) Larger ring size reduces error.

180°. This is true because the positron–electron combination will often be in motion during the process of annihilation, thereby altering the angle of ejection of the 511-keV photons. The detectors, however, assume a standard 180° emission path of the photons and therefore the localization of the positron emission is miscalculated in these cases (Fig. 8-12B).

Parallax Error

Resolution decreases toward the periphery of a ring of PET detectors. This is because some of the photons arising from peripheral annihilation events cross the ring of detectors at an oblique angle and can interact with one of several detectors along a relatively long path. When a photon interacts within a detector it is assumed the annihilation event occurred along a line of response originating at the front of the detector since the depth of interaction in the crystal is not recorded. The illustrations in Figure 8-13A show two possible lines of response from a single annihilation event occurring near the edge of the ring of detectors. This effect is sometimes referred to as a **parallax error** or **depth of interaction effect**.

The larger the size of the ring of detectors, relative to the size of the body being imaged, the less the effect, since the annihilation events will be more centrally located and the photons will cross the detector at a less oblique angle. This is illustrated in Figure 8-13B: the opposing photons from an annihilation event each cross two detectors in the smaller ring and only one in the larger ring.

Reconstruction and Post-Reconstruction Processing

Transverse slices are reconstructed using iterative reconstruction which is discussed in Chapter 7. Sagittal and coronal slices are constructed from transverse slices.

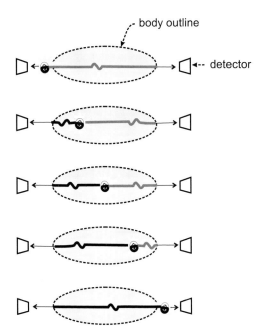

Figure 8-14 Attenuation is constant across a line connecting two detectors.

Attenuation in PET Imaging

Because of the relatively high energy of the positron-annihilation photon and because of the use of coincidence detection, attenuation correction is simpler and more accurate in PET than it is in SPECT. The attenuation coefficient for a 511-keV photon of PET is more nearly uniform across the various kinds of body tissues, i.e., fat, muscle, and bone, than it is for the lower energy photons encountered in SPECT. Also, the sum of the likelihoods of absorption for the two photons of the 511-keV pair will be the same regardless of the location of the annihilation event along the line of response. Figure 8-14 illustrates the paths of paired annihilation photons originating at different positions along the line. For example, for every position along this line-of-response, the length of the path traversed by the photon traveling to the right increases, the farther the annihilation has occurred from the point at which the photon exits. This increase is exactly offset by the decrease in the length of the path traversed by the left-traveling photon of the pair. In other words, adding the length of the path to the right to that for its pair-mate traveling to the left, one can see that the total amount of tissue traversed and therefore the total likelihood of absorption for the two photons of a pair will be the same no matter where along the line of response the annihilation occurred. If either photon of a pair is absorbed by surrounding tissue, the annihilation will not register as a coincident event and will not be counted.

Attenuation Correction

The attenuation in PET imaging, that is, loss of counts due to absorption of photons before they arrive at the detector, is compensated for arithmetically by using data from transmission scans. Depending on the camera design the transmission source can be a positron source, a high-energy single photon source, or a CT x-ray source. Each approach has unique advantages and disadvantages. In any case, the number of counts in each pixel with the patient in place is compared to that without the patient to determine for each pixel whether and by how much the count must be increased to compensate for the effect of absorption of photons in the patient's body.

Positron Emission Source

Using a positron source, such as ^{68}Ge, rotating around the patient (Fig. 8-15A), the count of events along the line of response for each detector pair with the patient in position (Fig. 8-15C) is compared to that without the patient (Fig 8-15B). The ratio of the counts without the patient in the gantry to the counts with the patient in the gantry for each line of response is used to calculate the correction factor for this detector pair when obtaining the emission scan (Fig. 8-15D).

Because the positron source rotates in a limited space between the patient and the inner wall of the camera (see Figure 8-15A), the source passes very close to one of the two opposing detectors in each pair. Due to this configuration the count-rate of this near detector may be very high. Because of the dead-time of the detector a significant portion

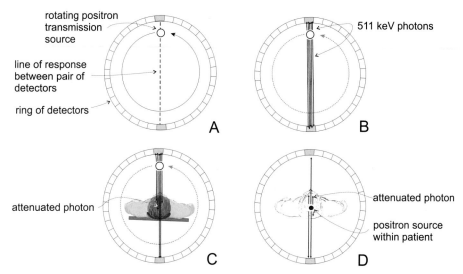

Figure 8-15 Attenuation correction using a positron transmission source. (A) Rotating source. (B) Transmission scan without patient. (C) Transmission scan with patient. (D) Emission scan. A sample detector pair is highlighted in grey.

of the photons emitted towards it will not be counted. This lengthens the time for acquisition of the attenuation correction data. One solution to the problem is to divide the transmission source into two or three less intense sources distributed around the camera ring to reduce the count rate for the near detectors.

High-Energy Single-Photon Emission Source
Another solution to the problem caused by allowing the transmission source to pass too close to one detector of a pair is to utilize only the distant detector of the pair. Of course when only one detector is used, coincidence is moot, and a relatively high-energy single-photon emitter such as ^{137}Cs (662 keV) can be used as the transmission source. The advantages of ^{137}Cs include its lower cost and its longer half-life compared to a ^{68}Ge positron source (30 years compared to 280 days). As a result a higher activity can be used since camera dead time from a close detector is no longer a factor. Shorter acquisition times are possible since a source of higher activity can be used.

A disadvantage of the single-photon technique is that scatter photons are a greater problem than with coincidence-detection imaging. Also, the attenuation coefficients, calculated from measurements made with the 662-keV photons of ^{137}Cs, underestimate the attenuation of the lower energy 511-keV annihilation photons.

Computed Tomography X-Ray Source
Transmission attenuation data can be obtained using a rotating x-ray source. PET-CT cameras may be configured to use their x-ray data for calculating attenuation correction for PET emission scans. X-ray photon energies are generally less than 140-keV, markedly lower than the 511-keV annihilation photons. The linear attenuation coefficients measured with the x-ray must be scaled down to the values for 511-keV photons prior to being used to correct the emission data.

At the photon energies of x-rays, the attenuation coefficients for bone and soft tissue differ significantly from each other. (The difference is much less significant for the 511-keV photon than for x-rays.) The scaling factors required to "adjust" the attenuation coefficients for x-rays to the attenuation coefficients for 511-keV photons must take into account the differences between bone and soft tissue. To allow for this, the CT

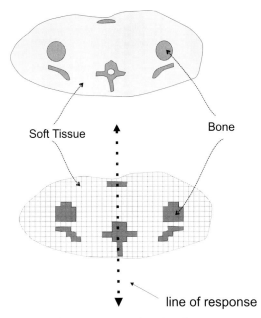

Soft Tissue

Bone

line of response

Figure 8-16 Segmentation of CT data for attenuation correction.

attenuation map is **segmented** into those pixels that correspond to bone and those that correspond to soft tissue (see Fig. 8-16). Once the appropriate scaling factors are applied to each pixel of the segmented map the attenuation along each line of response is calculated and used for the attenuation correction of the positron emission scans.

Standard Uptake Values

Determination of the amount of an injected radionuclide that is taken up by a tumor or organ is used in ^{18}F-fluoro-deoxy-glucose (FDG) scans to aid in the differentiation of benign from malignant masses. Because accurate measurement of uptake is confounded by, among other things, the variable and often poorly known extent of photon absorption in the surrounding tissues, a semi-quantitative measure, the **standard uptake value (SUV)**, is commonly used as an estimate of the actual uptake.

Calculation of the SUV requires a PET estimate of the concentration of activity in the tumor or organ in mCi/ml or mBq/ml, the mass of the patient in grams, and the amount of the injected activity in mCi or mBq.

SUV = (concentration of activity in tumor

in mBq/ml) × (mass of patient in g)

× (injected activity in mBq)$^{-1}$

The units of SUV are therefore g/mL, but since tissue is almost completely water, and 1 ml of water weighs 1 g, SUVs are given without units.

SUV measurements are overestimated by the above formula in patients who are overweight because fat does not concentrate FDG as much as the rest of the tissue of the body. The contribution of the patient's weight towards the SUV is reduced if the patients body surface area (BSA) is used in the calculation instead of his mass. This is because the formula for the calculation of the BSA incorporates the patient's height as well as his weight. BSA is estimated by [2]

Body surface area = (weight in kg)$^{0.425}$

× (height in cm)$^{0.725}$ × 0.007184

The formula for SUV based on body surface area becomes [3]

SUV_{BSA}

= (calculated activity in the region of interest)

× (body surface area)/(injected activity)

References

1 Llewellen, T and Karp, J, Pet Systems, pp. 179–194. In Wernick, MN and Aarsvold, Jn, *Emission Tomography, The Fundamentals of PET and SPECT*, New York, Elsevier, Inc., 2004, p. 180.

2 DuBois, D and DuBois, EF, A Formula to Estimate the Approximate Surface Area if Height and Weight be Known. *Arch Int Med*, 1916, 17: 863–871.

3 Kim, CK, Gupta, NC, Chandramouli, B, and Alavi, A, Standardized Uptake Values of FDG: Body Surface Area Correction is Preferable

to Body Weight Correction. *Journal of Nuclear Medicine*, 1994, 35: 164–167.

Test Yourself

1 Match the following phrases with the choices listed below the phrases:

(a) Two 511 keV photons arising from a single annihilation event striking opposing detectors simultaneously.

(b) Two 511 keV photons arising from separate annihilation events striking opposing detectors simultaneously.

(c) One 511 keV photon from a single annihilation event strikes a detector, the other is absorbed by surrounding tissue.

Choices: Singles event, Random event, True coincidence event.

2 Which of the following characteristics are desirable for the crystalline materials used to detect 511 keV photons:

(a) Low density

(b) High atomic number

(c) Long decay time

(d) High light yield.

3 True or false: Coincidence circuitry can more readily differentiate singles events from true coincidence events than random events from true coincidence events.

4 Which of the following factors tend to reduce resolution in PET imaging:

(a) The positrons can travel a significant distance prior to annihilation.

(b) Septa in 3-D imaging decrease the number of detected true coincidence events.

(c) 511 keV photon emissions are not always exactly 180° apart following an annihilation reaction.

(d) Proximity of source to edge of ring of detectors.

5 Attenuation correction can be performed in PET scanning using which of the following rotating transmission sources:

(a) A positron source such as ^{68}Ge

(b) A rotating x-ray source

(c) High energy single-photon emission source such as ^{137}Cs

(d) All of the above.

6 True or false: Standard uptake values (SUV) measurements based on body surface area may be more accurate than those based on weight alone in obese patients because adipose tissue does not concentrate FDG as much as the rest of the tissue in the body.

9 Combined PET/CT imaging

For specific clinical diagnoses positron emission tomography (PET) can detect more sites of disease than conventional anatomical imaging such as **x-ray computed tomography (CT)** or magnetic resonance imaging (MRI). Interpretation of PET can be difficult however because PET images have few anatomical landmarks for determining the location of abnormal findings. Combining PET and CT images acquired sequentially on their separate devices provides a partial solution to this problem. Unless patient positioning is carefully reproduced between the studies, the PET and x-ray images will not match, and the resulting misregistration will lead to inaccuracies in determining the anatomic location of the abnormalities seen in the PET images. In one approach to correcting for positioning errors, the images from a PET camera and a CT scanner are fused manually or by computer software or both. To further ensure accurate registration of the PET and CT data the two imaging modalities can be physically combined in a single unit, the **PET-CT camera.** This combination unit also facilitates the use of the CT data to correct the PET images for the attenuation of the annihilation radiation as it traverses body tissues en route to the detector.

In this chapter we will briefly introduce x-ray tubes and how they produce x-rays. We will then discuss the general configuration of a CT scanner. This will be followed by an illustration of a PET-CT camera.

X-Ray Production

Wilhelm Roentgen is usually credited with discovering x-rays. Although they had been observed earlier, Roentgen was the first to describe their basic properties in 1895, and in 1901 he was awarded the first Nobel Prize in physics. When he first observed them, the nature of the radiations was unknown and their properties surprising, if not actually mysterious. This is reflected in his use of the "x" in naming them "**x-ray**."

A modern **x-ray tube** (Fig. 9-1) is an evacuated glass or ceramic tube with a window for exiting the x-rays. Within the vacuum tube, electrons are "boiled off" an electrically heated filament wire, the cathode, and are accelerated to high speed toward the positively-charged tungsten **target**, the anode, by the high potential difference (or voltage) maintained between filament and target (Fig. 9-2). The vast majority of these electrons interact with outer-shell electrons of the tungsten target and their kinetic energy is lost as heat. A small percent of the electrons bombarding the target, approximately 0.2% of them, cause the emission of an x-ray by either characteristic radiation or bremsstrahlung (see Chapter 2).

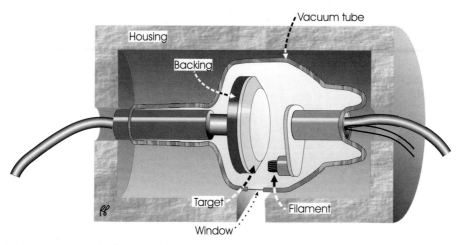

Figure 9-1 Basic components of an x-ray tube.

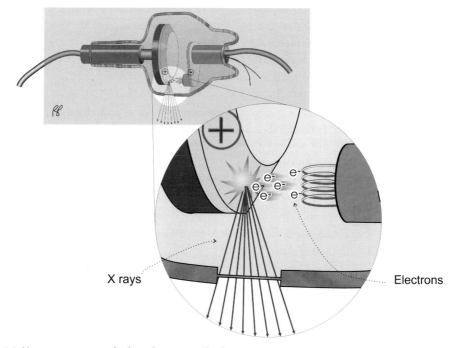

Figure 9-2 X-rays are generated when electrons strike the tungsten target.

Medical imaging utilizes both characteristic and bremsstrahlung x-rays. Although characteristic x-rays are discrete, bremsstrahlung interactions produce a full range of x-ray energies, from 0 keV to a maximum which is defined by the **maximum** or **peak voltage** applied between the filament and target. The larger this peak applied voltage (expressed in kilovolts and abbreviated kV_p) the higher the maximum x-ray energy and the greater the number of x-rays created. The lower energy x-rays in the x-ray beam add to the radiation dose to the patient but not to the

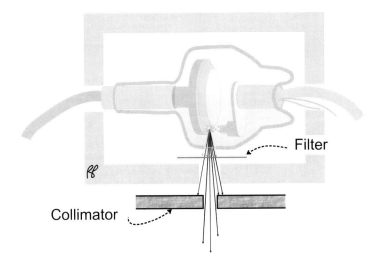

Filter

Collimator

Figure 9-3 The filter attenuates the lower energy x-rays (depicted as lighter grey arrows) and a collimator narrows the x-ray beam.

quality of the images, and so these are attenuated by a **filter** (usually aluminum) placed in front of the tube window (Fig. 9-3). Thus both the applied voltage between the filament and target and the attenuation by the filter affect the range of energies of the emitted x-rays. In addition, the amount of electricity or current (expressed in milliamperes or **mA**) used to heat the filament to boil off the electrons and the duration of this current (in seconds) affects the total number of x-rays emitted. The last two quantities are often combined as a product and referred to as **mAs** (milliamperes-seconds). Finally, the x-ray beam is narrowed or collimated by lead shutters (Fig. 9-3) to avoid radiating tissues other than those being imaged.

An unavoidable accompaniment of bombardment of the target is the generation of **heat**. The tungsten, itself, is dense enough to stop the electrons and capable of withstanding the heating. The heat is dissipated from the tungsten by a copper backing and by rapid rotation of the target, which spreads the heat over a larger area and allows a brief rest period for any one spot to "cool down" before serving again as the focus point of the electron beam. Further methods of cooling include limiting the duration of the bombardment, and external air or oil cooling.

The contents of the x-ray tube are in a vacuum so that there are no gas particles to impede the flow of the electrons.

X-Ray Imaging

Images created by projecting x-rays at the body, also called x-rays by convention, are in essence shadows. They can be thought of as "inverse shadows"; the denser tissues such as bone, which block much of the x-rays, appear as white on the film or digital detectors, and lungs, which are mainly composed of air, appear to be very dark in the image.

Computed Tomography

Overview

When a nuclear medicine gamma camera head is stationary, static or planar images are obtained from the gamma emissions emanating from a patient. If the camera head is rotated mechanically in small increments around the same patient and multiple images (called projections) are collected and digitized, this data can be "back-projected" to create transaxial slices. This process of acquisition and reconstruction is called SPECT (single photon computed emission

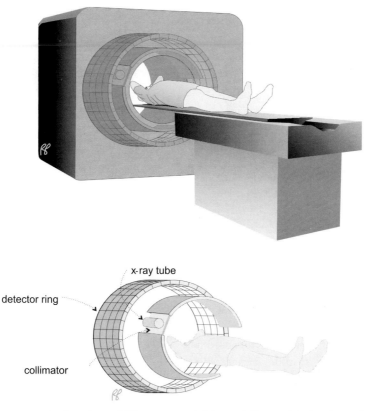

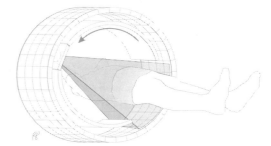

Figure 9-4 Basic components of one type of CT scanner containing a stationary detector ring and rotating x-ray tube.

tomography) and was covered in depth in Chapter 6.

In a similar manner an x-ray of a patient using a stationary x-ray source and detector is called a planar image. Chest x-rays are probably the most common example of a planar x-ray image. If, on the other hand, the x-ray data is recorded over the full 360° path encircling the patient this data can also be "back-projected" to create transaxial slices. The x-ray source and detectors in most current scanners are arranged in one of two configurations. Either the x-ray source rotates within a stationary complete ring of detectors (called **rotate-stationary** systems) as illustrated in Figures 9-4 and 9-5 or the x-ray source and an opposing arc of detectors rotate in synchrony around the patient (called **rotate-rotate** systems) as seen in Figure 9-6. This process of acquisition and reconstruction is called

Figure 9-5 Rotate-stationary configuration. A rotating source and collimator generate a fan shaped x-ray beam which is directed towards a stationary ring of detectors.

Computed (trans)Axial Tomography (CAT) or simply **Computed Tomography (CT)** scanning.

The x-ray source is moved in increments around the patient. At each position the x-ray tube is turned on and the patient is exposed to a fan-shaped beam of x-rays (Fig. 9-5).

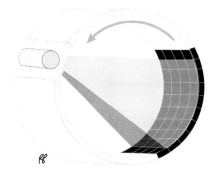

Figure 9-6 Rotate-rotate configuration. Opposing source and detector rotate synchronously.

The fan beam is shaped by a collimator placed between the tube and the patient. The x-rays that are not attenuated by the patient's body are registered by the detectors on the opposite side of the patient. The detectors are composed of **ceramic scintillators** which like the NaI(Tl) crystal discussed in Chapter 5 emit light in response to x-rays. Because the scintillator detectors used in CT scanners must respond to the large rapidly changing flow of x-rays generated by the CT x-ray source, the chemical composition of the materials in the ceramic are more complex than the NaI(Tl) crystal. In particular these scintillators must have very rapid decay times; both the initial light output in response to the x-ray excitation must be rapid, and the residual light within the scintillator present after the initial response, called the **afterglow,** must dissipate rapidly.

The ceramic scintillators are backed by **photodiodes** which generate electrical pulses/(current) in response to the light photons. The photodiodes are semiconductors that function similar to photomultiplier tubes (PMTs) by converting photon energy into current. Semiconductors and PMTs are discussed in Chapters 4 and 5 respectively.

Axial and Helical Scanning

Older CT scanners acquired individual axial slices. The patient bed (pallet) was advanced in small increments. After each increment the bed was stopped and an axial slice was acquired (Fig. 9-7A).

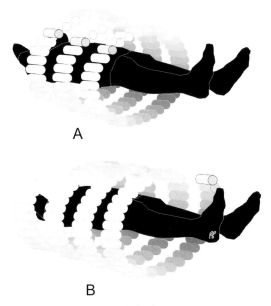

A

B

Figure 9-7 Axial versus helical scanning.

Newer CT scanners acquire continuously as the bed advances continuously so that the path of the x-ray source is much like the peel of a tubular apple or a helix, and therefore this is referred to as a **helical** scan (Fig. 9-7B). Some authors use the name **spiral** instead of helical to refer to the same process.

There are two advantages to helical scanning: the first is the shorter duration of the study due to faster scanning times, and the second is the increased flexibility during data reconstruction. The angle of reconstruction of the axial slices can be chosen by the operator and higher quality coronal and sagittal slices are created from these datasets.

Pitch

The helical motion of the gantry of the scanner can be described by specifying both the rotational speed of the gantry in revolutions per second and the distance the patient bed travels along the long axis of the patient in millimeters for each revolution of the gantry. The latter is called the pitch of the helix, an engineering term for describing the threads of a screw or bolt. The uppermost

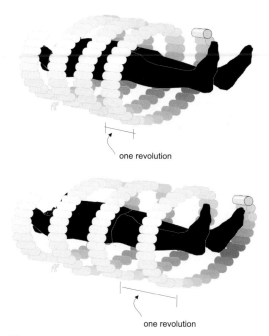

one revolution

one revolution

Figure 9-8 Pitch.

illustration in Figure 9-8 shows an acquisition with a shorter distance traveled per revolution, or a smaller pitch, than the one shown in the lower illustration.

There are many technical definitions of the term **pitch** when applied to helical CT. The most commonly used definition is the distance the table advances in one complete revolution divided by either the slice width or the width of the collimated x-ray beam. In a single slice helical CT scanner these values of pitch are equivalent since the slice width is the width of the collimated x-ray beam. In a mutlislice scanner several slices are acquired with each sweep of the x-ray beam, so the two different values for pitch can be calculated depending on whether the slice width or the width of the beam is used in the denominator of the equation.

Hounsfield Units

CT pixel intensities are given in CT numbers or **Hounsfield unit** (HU) and are simply scaled units of attenuation as measured by CT. The Hounsfield unit is named for

Sir Godfrey Hounsfield who developed the first practical CAT scanner and who, along with Allan Cormack, was awarded the 1979 Nobel Prize in Medicine. It is not easily converted to SI units; nevertheless, radiology uses the Hounsfield unit for dose calculation in preference to the attenuation coefficient used by most other disciplines. If μ is the average linear attenuation coefficient for the pixel of interest and μ_w is the value for water, then the CT number of HU is given by

$$HU = (\mu - \mu_w)/\mu_w$$

Tables of Hounsfield units are available. Air, which stops virtually no x-radiation has a value of -1000 HU; water, which moderately attenuates the x-ray beam, has a value of 0 (zero) HU, and for bone, which blocks a large fraction of the beam, has a value of 1000 HU, the highest available number on the Hounsfield scale. The HU for fat is about -10 while the firmer of the soft body tissues have HU values in the range from about 10 to 60.

A more in-depth discussion of photon attenuation is covered in Chapter 2.

PET-CT Cameras

PET-CT cameras are presently configured as sequential gantries, also called in-line cameras, with a shared patient bed, or pallet (see Fig. 9-9). The advantage of this configuration as discussed at the introduction of the chapter is the consistent positioning of the patient between the acquisitions which reduces the risk of misregistration of the images. In addition the CT data can be used for attenuation correction of the PET images. Attenuation correction using CT data is discussed in Chapter 8.

The CT scanner is usually placed closer to the patient. The CT scan is acquired in its entirety prior to acquiring the PET scan (Fig. 9-10), although the order of acquisition of the studies can be reversed.

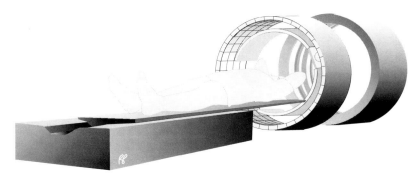

Figure 9-9 PET-CT.

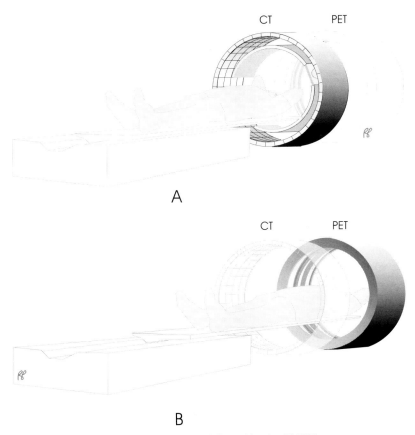

Figure 9-10 PET-CT. (A) The entire CT scan is acquired, followed by the (B) PET scan.

Current Limitations in PET-CT Imaging

Breathing Artifacts
PET-CT cameras have mitigated the majority of the PET and CT registration problems caused by differences in patient positioning which occur when the patient must be physically moved between independent PET and CT units. However, due to differences in breathing patterns between CT, where breath holding is desirable, and PET imaging, where breath holding is not possible, misalignment, particularly

near the diaphragm, can cause misregistration of images in the lower lungs and upper abdomen. In addition, if the CT data is used for attenuation correction, artifacts may be introduced into the PET images in the same areas. To assure better alignment of the diaphragm some institutions instruct patients to breathe normally, or shallowly, during both CT and PET acquisitions.

Contrast Agent Artifacts

The use of intravenous and oral contrast during CT imaging can improve anatomic localization, however, contrast is relatively dense and can alter the attenuation maps that are constructed from CT data. In particular the x-ray attenuation will be greatly increased at sites of greater concentration of contrast, such as pooling of the oral contrast in the colon, or vascular filling with an intravenous bolus. Although the 511 keV photons are also attenuated by contrast, the distribution of the contrast can change between the time of acquisition of the CT and PET studies. As a result the CT attenuation map may not correctly approximate the PET photon attenuation in these specific areas.

In addition the HU, or amount of x-ray attenuation, will also be somewhat increased in the soft tissues into which the contrast diffuses. Therefore the attenuation coefficient scaling factors (see Chapter 8, section on attenuation correction) for soft tissue which are based on non-contrast CT x-ray attenuation are not as accurate. Therefore, for the above reasons, use of attenuation correction data from contrast enhanced CT studies to correct attenuation in PET studies may result in artifacts in the PET images.

Test Yourself

1 True or false: X-ray production is an efficient process. The majority of electrons striking the tungsten target cause the emission of x-rays.

2 Which of the following statements are true about the x-ray beam produced by x-ray tubes?
  (a) The x-rays are monoenergetic; they all have the same keV
  (b) The beam is a combination of characteristic and bremsstrahlung radiation
  (c) The maximum x-ray energy is not affected by the peak voltage applied between the filament and target
  (d) The number of x-rays produced in the target increases as the current in the filament increases.

3 Which of the following statements are true concerning CT scanning
  (a) CT images acquired in conjunction with PET images can provide useful correlative anatomic information
  (b) Helical CT scanning allows for faster acquisition times and greater flexibility in image reconstruction compared to conventional axial CT scanning
  (c) Current CT scanners consist of a stationary ring of detectors with an inner rotating x-ray source or a rotating arc of detectors that oppose the rotating x-ray source
  (d) All of the above.

4 True or false: A common definition of the term pitch, when used in reference to helical CT scanning, is the ratio of the distance the patient bed advances per gantry revolution divided by the width of the collimated x-ray beam.

5 Select the Hounsfield units (HU) from the values below that most closely correspond to the following:
  (a) Bone
  (b) Fat
  (c) Muscle
  (d) Air
  (e) Water.
  HU values: -1000, -10, 0, 30, 1000.

10 Quality control

To ensure dependable performance of equipment, each nuclear medicine department is required to perform a routine series of tests on each device. These tests comprise the quality control program for the department.

Nonimaging Devices

Dose Calibrator

The testing performed on dose calibrators is quite rigorous to ensure that correct radiopharmaceutical doses are administered to patients. Dose calibrators are checked for accuracy, constancy, linearity, and geometry.

Accuracy

Accuracy is a measure of the readings of the dose calibrator in comparison to well-accepted standards. Two long-lived nuclide sources, such as ^{137}Cs (half-life = 30 years) and ^{57}Co (half-life = 270 days), are measured repeatedly in the calibrator and the average readings are compared to values issued by the National Institute of Standards and Technology. If the reading differs from the standards by more than 10%, the dose calibrator should not be used. Accuracy should be checked at *installation, annually*, and *after repairs* or *when the instrument is moved* to a new location within the clinic.

Constancy

To ensure that the calibrator readings are constant from day to day, a long-lived nuclide such as ^{137}Cs is measured in the dose calibrator. The reading should not vary by more than 10% from the value recorded at the initial accuracy test corrected for the decay of the ^{137}Cs standard. The readings at each predefined setting (or "channel")—^{99m}Tc, ^{131}I, ^{67}Ga—are recorded and compared to previous readings. **Constancy** (including the channel checks) is performed *daily*.

Linearity

Linearity tests the calibrator over the range of doses used, from the highest dose administered to a patient, down to 10 μCi. One method of checking linearity is to measure the maximum activity that a department will use, such as 7.4 GBq (200 mCi) of ^{99m}Tc, and to repeat the measurement at intervals of 6, 24, 30, 48, and 96 hours. A more rapid technique is to measure the dose unshielded, then repeat the measurement of the same dose shielded within **lead sleeves** of varying thickness. The thickness of the sleeves is such that when used both individually and in combination they effectively reproduce the decline in activity of the ^{99m}Tc seen over 96 hours (Fig. 10-1). The sleeves should be carefully examined prior to use; they will not yield accurate readings if there are any cracks or dents. The initial linearity check should be performed with the slower method of measuring a sample as it decays. Linearity should be checked at *installation* of the device, *quarterly*, and *after repairs* or *when the instrument is moved* to a new location within the clinic.

Figure 10-1 Linearity sleeves.

Geometry

The apparent activity of a dose will vary with the volume and shape of the container and the position of the dose within the chamber. The effect of sample **geometry** can be tested by placing a small amount of activity at the bottom of a container and progressively adding a diluent such as water. Similarly, a sample can be measured in each of the types of containers—vials, syringes, bottles, and so on—used within the laboratory. The left side of Figure 10-2 illustrates photon absorption in a relatively large volume of diluent; the right side illustrates photon loss through the calibrator opening when the sample is placed in a taller container. Dose measurements should not vary by more than 10%. The effect of sample geometry should be checked at *installation* and *after repairs*.

Survey Meters

Constancy

A long-lived radioactive source should be checked *daily* to ensure that the meter reading of the source is constant (within 10% of its original value). If the reading differs by more than 10% of the original value, the survey meter should be

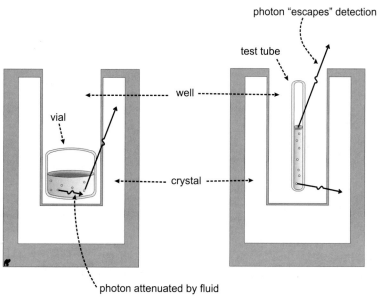

Figure 10-2 Variation in sample geometry.

recalibrated. Typically the source is attached to the side of the survey meter.

Calibration

All survey meters, including Geiger counters, must be checked for accuracy. To do so, readings are taken from two long-lived sources (^{57}Co and ^{137}Cs) at incremental distances (such as every tenth of a meter up to one meter) from the sources. The sources should be radioactive enough to generate readings at approximately one-third and two-thirds of full scale. The readings must be within 20% of their expected measurement. This test should be performed at *installation*, *after repairs*, and *annually*.

Crystal Scintillation Detectors: Well Counters and Thyroid Probes

Calibration

On a *daily* basis or prior to each patient measurement the photopeak is checked using a long-lived nuclide reference source (usually ^{137}Cs) which is placed in the well counter or in front of the probe. The voltage or gain settings are adjusted (see Chapter 5) manually or automatically for the maximum peak (count rate); the settings and the peak count rate should be recorded. The count rate should not differ from the previous average values or a similarly established value by more than 10%. The full-width half-maximum (FWHM) of the photopeak should be <10% of the energy of the photopeak. Calibration should be routinely performed as well as performed *annually* and *following repairs*.

Efficiency

This *annual* test is performed using reference standard sources with similar energy emissions to the nuclides measured routinely in the counter or in front of the thyroid probe. For example for the thyroid probe, barium-133 (prinicipal gamma emission 356 keV) can be used as a reference for iodine-131 (364 keV), and cobalt-57 (122 keV) can be used for iodine-123 (157 keV).

The **efficiency** of the thyroid probe or well counter can be calculated using the following equation:

Efficiency
$$= [\text{(counts per minute of standard)} - \text{(counts per minute of background)}] \times [\text{(activity of standard in } \mu\text{Ci)}]^{-1} \text{ [1]}$$

Imaging

Planar Gamma Camera

Photopeak

Prior to imaging, the pulse-height analyzer may require adjustment to properly center the window of photon energies accepted. The procedure of adjustment differs from camera to camera.

One simple procedure for checking the location of the energy window is to place a vial or syringe containing a small quantity of isotope against the collimator. The computer screen displays a plot of counts vs. energy and the current location of the window (see Fig. 5-6). The user can then adjust the location and width of the window. For example, a standard setting for technetium-99m is a photopeak of 140 keV and a window of 20%. The window is set between 126 keV and 154 keV. A narrower window rejects more scatter photons but also reduces the counts from the patient. This increases the resolution but decreases the sensitivity of the camera.

Drift of the energy windows away from the peak will lead to significant artifacts in images. Off-center windows will yield relatively "hot" or "cold" photomultiplier defects on the daily uniformity floods. Figure 10-3 presents images of a uniform source acquired with varying energy windows. In the upper images the energy window is centered above the photopeak; in the lower images the window is centered below the photopeak.

Uniformity Floods

Ideally, a scintillation camera should produce a uniform image of a uniform source.

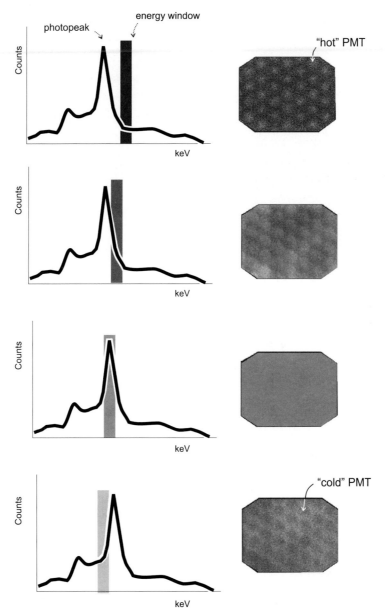

Figure 10-3 Nonuniformity due to off-center energy windows (images courtesy of Philip Livingston, CNMT).

Unfortunately this ideal is not met due to imperfections in the collimators, variations in crystal response, differences among photomultiplier tubes' response, and minor fluctuations in the electrical circuitry.

The uniformity of the camera's response can be checked by imaging a flood source. A solid plastic disk manufactured with 5 to 20 mCi of ^{57}Co uniformly distributed throughout its extent (or a fluid-filled sheet source containing a dilute solution of radioactivity) is placed directly on the collimator, the so-called **extrinsic flood field**. Alternatively, a point source is suspended several feet (at a distance three to four times the diameter of the crystal) directly above the surface of the crystal from which the collimator

Figure 10-4 Uniform flood field.

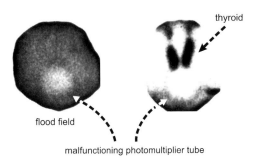

thyroid

flood field

malfunctioning photomultiplier tube

Figure 10-5 Defective photomultiplier tube.

has been removed, the so-called **intrinsic flood field**. A 5- to 30-million-count image of the flood is then collected according to the manufacturer's directions.

Daily Visual Inspection for Marked Nonuniformity
Images of the flood source should be collected and examined *daily*. The human eye can discern variation in counts that are as small as 15%. A uniform flood image is shown in Figure 10-4. If the image does not appear to be uniform, corrective action is required. An example of nonuniformity is shown in Figure 10-5 where a large, rounded defect suggests malfunction of a photomultiplier tube. This corresponds to a similar defect in an image of the thyroid. Less common, but more serious, is the sharply demarcated defect caused by a crack in the crystal (Fig. 10-6).

crack in crystal

Figure 10-6 Cracked crystal.

These defects are detectable with or without the collimator in place. A subtler example of nonuniformity, from slow drift in the energy window circuitry over a period of three weeks, is shown in Figure 10-7. A damaged collimator can cause linear photopenic defects.

Correction of Nonuniformity
In addition to the visual inspection, a flood source collected without collimation (intrinsic) is imaged and loaded into the computer (weekly for older camera systems, much less frequently for newer systems). The computer identifies regions in which counts differ significantly from the mean counts. To compensate for these differences, the computer determines a correction for each pixel, which will later be applied when the camera is used to image a patient. The correction data stored in the computer are referred as the **uniformity correction matrix**. Figure 10-8 is a simplistic representation of the application of a uniformity correction matrix.

Spatial Resolution
Bar Phantom
The resolution of the imaging system is evaluated visually by imaging a **bar phantom**. One type of bar phantom is made of lead bars encased in lucite. The phantom is divided into four quadrants, and the bars within each quadrant are

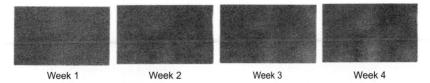

Week 1   Week 2   Week 3   Week 4

Figure 10-7 Nonuniformity due to a drift in circuitry.

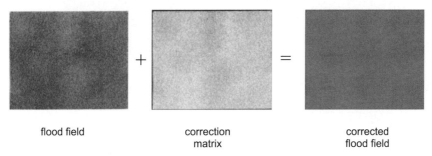

flood field   correction   corrected
                            matrix   flood field

Figure 10-8 Uniformity correction matrix.

spaced in regular intervals; this interval is different in each quadrant. The intervals illustrated in Figure 10-9 are hypothetical but representative of a standard configuration.

The flood source is imaged with the bar phantom placed between it and the collimator. The quadrant with the closest bar spacing that can be distinguished is a measurement of the resolution of the system. The spatial resolution is checked *weekly* for degradation of resolution.

The bar phantom can be used to demonstrate that spatial resolution decreases as the distance between the source and the camera increases, as shown in Figure 10-10. In the images on the left, the bar phantom is placed directly on the collimator; the minimum resolvable bar spacing is 4.0 mm. On the right, the phantom is positioned 10 cm away from the collimator, and the minimum resolvable bar spacing is 4.8 mm.

Line Spread Function
At installation and following repair, the spatial resolution of the system is checked using the **line spread function**. This can be accomplished by placing a lead mask containing slits (NEMA line pattern) over the camera face, once the collimator has been removed, and suspending a point

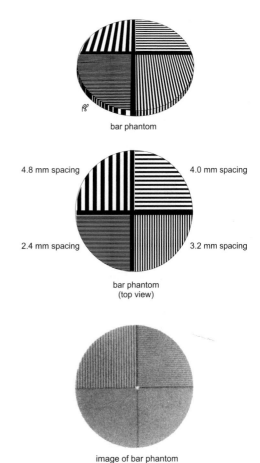

bar phantom

4.8 mm spacing   4.0 mm spacing

2.4 mm spacing   3.2 mm spacing

bar phantom
(top view)

image of bar phantom

Figure 10-9 Bar phantom.

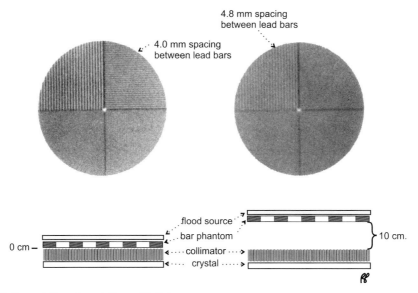

Figure 10-10 Degradation of resolution with distance.

source several feet above the lead mask to create a series of lines on the crystal. This is an intrinsic test as it is performed without the collimator on the crystal. With the collimator in place, an extrinsic test of the line spread function can be performed with a line source of radioactivity placed on the surface of the collimator. In either case, the image of the line is broader and less distinct than the slit or line source for a variety of reasons, including acceptance of angled photons, scatter in the material around the source and in the crystal, and imperfections in the camera circuitry (Fig. 10-11). Figure 10-12 contains a plot of the distribution of counts in a horizontal slice through the image of the source (the breadth is dependent on the characteristics of the photon and the imaging system). The **full-width at half-maximum (FWHM)** and full-width at tenth-maximum (FWTM) are measurements of the spread of this curve (thus the name line spread function). The FWHM is the width of the curve at half the peak count, likewise the FWTM is the width of the curve at one-tenth of the maximum counts (see Fig. 10-12). A system with better resolution will have a narrower curve and therefore smaller FWHM and FWTM.

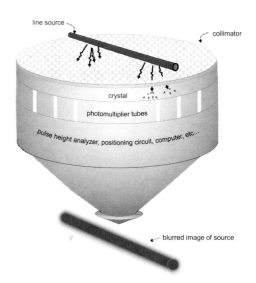

Figure 10-11 Blurring of a line source.

Linearity

Linearity of the gamma camera image is tested by examining the image of the bar phantom obtained with a high-resolution collimator in place. The lines in the image should be straight and unbroken. Note that linearity here refers to the appearance of the bars as lines, whereas the same word in reference to the dose calibrator refers to the relationship between dose and meter reading.

MODULATION TRANSFER FUNCTION

An alternative presentation of these data is achieved by expressing the line spread function in the frequency domain. This is an application of the Fourier transformation described in Chapter 7. As we discussed before, images can be represented as a collection of waves in frequency space. The **modulation transfer function** plots show the ability of the camera to reproduce these waves at each of the spatial frequencies. Each camera system has a unique frequency response. This is analogous to stereo systems; some reproduce more of the treble (higher frequency waves), others reproduce more of the bass (lower frequency waves), and the better the system the more of both bass and treble it can reproduce.

In nuclear medicine systems the higher frequencies are necessary to reproduce the sharp edges and fine details of an image; the lower and middle frequencies produce the remainder of the image. Figure 10-13 contains plots of the modulation transfer functions of two hypothetical camera systems. System A accepts a greater proportion of lower frequencies, while system B accepts a greater proportion of higher frequencies.

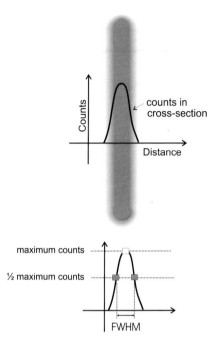

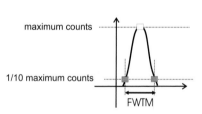

Figure 10-12 Full-width at half-maximum and full-width at tenth-maximum.

SPECT

The following additional quality control procedures are necessary for SPECT cameras. The frequency of performance of these tests will vary according to the manufacturer's recommendations.

Uniformity

SPECT images are degraded by small degrees of nonuniformity in the flood field that do not adversely affect planar images. It is important to acquire the **flood uniformity correction** for approximately 100 million counts (depending on manufacturer's recommendations) to reduce nonuniformity caused by statistical variations in count rate. During backprojection, relatively minor defects will become quite prominent and sometimes appear as ring artifacts in the reconstructed transaxial slices. The camera head

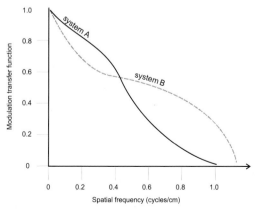

Figure 10-13 Modulation transfer function.

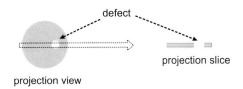

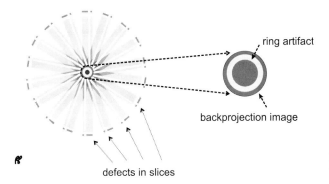

Figure 10-14 Ring artifact created during backprojection of an area of nonuniformity.

effectively "drags" the defect with it as it circles the patient. Figure 10-14 illustrates this effect.

Center of Rotation

It is assumed that the camera heads will rotate in a near perfect circle (or ellipse) and that heads will remain almost precisely aligned in their opposing positions. It is also assumed that the predicted or "electronic" center of the path of rotation will match the "mechanical" or actual center of the camera head rotation. Deviation from either expectation will degrade image resolution and can be seen as a displacement of the **center of rotation (COR)**. Probably the most common cause of apparent displacement of COR is a result of inadvertent errors (not leveling the camera head, bumping the table, and so on) during data collection. The most common cause of true shift of the COR is electronic malfunction. Mechanical problems, such as the use of a collimator that is too heavy for the gantry, are less common. Figure 10-15A illustrates a near perfect circular rotation path for a camera. The predicted COR (cross) is aligned with the actual

(mechanical) COR (circle); Figure 10-15B illustrates an elliptical orbit caused by a heavy camera head that at the bottom of its orbit drifts downward under the influence of gravity. The center of this elliptical orbit (circle) is offset from the center of the predicted circular orbit (cross).

Measurement of COR

The test for a stable COR consists of placing a point source slightly off center on the patient bed. Projection views are obtained over a 360° arc (upper panel of Fig. 10-16).

There are several ways to analyze the difference between the actual and predicted COR. One common approach is to plot the position of the point source as a function of camera head position along the circle of rotation and compare this to the predicted values. The location of the point source is plotted in the x (perpendicular to the long axis of the bed) and y (parallel to the long axis of the bed) directions. The position of the source plotted in the x direction should closely approximate

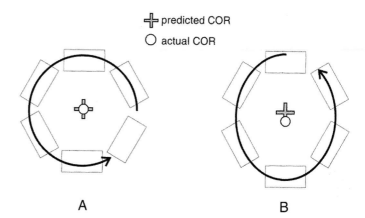

Figure 10-15 Deviation of the mechanical COR.

A   B

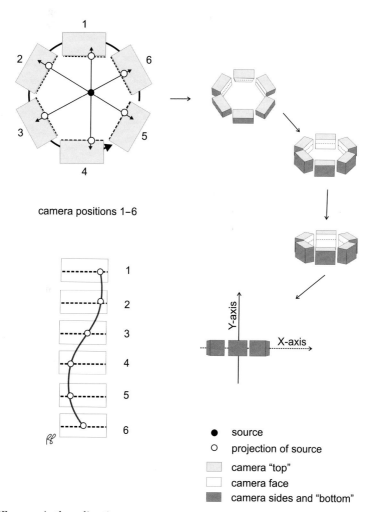

camera positions 1–6

Figure 10-16 COR curves in the x direction.

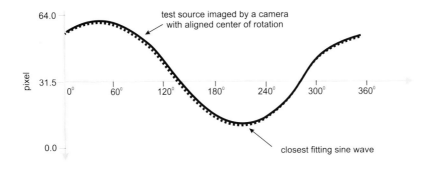

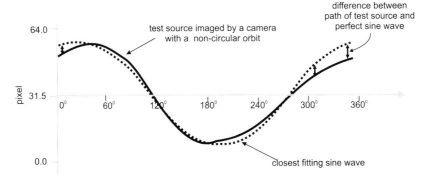

Figure 10-17 Normal and abnormal COR tests.

a sine wave (lower panel of Fig. 10-16). The plot of the image of the source in the y direction (not depicted) should be a straight line. The plot of the source on a camera with a normal COR is illustrated in the upper portion of Figure 10-17. The bottom graph plots the same source in the camera with the elliptical orbit depicted in Figure 10-15. This plot deviates by several pixels from the expected sine wave (dashed lines). Along a portion of the rotation, the measured position differs by several pixels from the expected position of the source. This COR test is abnormal. Deviations of greater than one half pixel from the expected position of the source are considered abnormal and should be checked with a second collection. A persistent abnormality will require repair prior to collection of further SPECT studies. The results of the COR calculation are stored in the computer and used to correct subsequent SPECT scans.

Measuring SPECT Uniformity and Resolution with a Phantom

Phantoms are fillable objects that are used to assess camera resolution. A phantom can be as simple as a plastic bag or as complex as a model of a slice of the brain. SPECT phantoms are cylindrical lucite containers in which are interspersed different sized rods, cylinders, and/or cones of lucite. The container is then filled with water containing a small amount of radioactivity. Tomographic images of the phantom are collected and the reconstructed slices are inspected for visibility of the lucite objects. The images should be compared to prior acquisitions for evidence of degradation in resolution.

PET

The quality-control protocols for PET cameras can be quite extensive and vary among

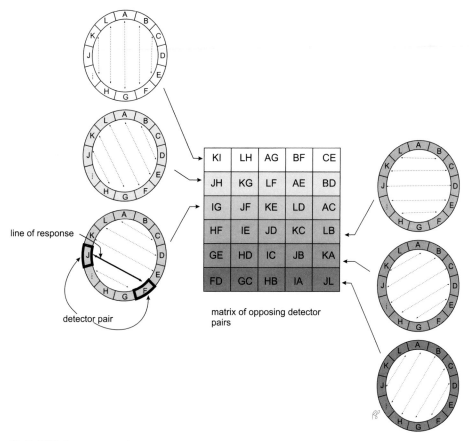

KI	LH	AG	BF	CE
JH	KG	LF	AE	BD
IG	JF	KE	LD	AC
HF	IE	JD	KC	LB
GE	HD	IC	JB	KA
FD	GC	HB	IA	JL

line of response

detector pair

matrix of opposing detector pairs

Figure 10-18 PET sinogram matrix.

manufacturers. A few of the more routine procedures are summarized briefly in this section as an introduction to this topic for the reader.

Blank Transmission Scans

A common *daily* quality control procedure is the acquisition and evaluation of a scan using the internal transmission source or a separate low activity source, without a patient in the field of view. Because there is no patient present this is called a **blank scan**. The blank scan is examined for evidence of defective detectors. The data acquired from the blank scan can be displayed as a sinogram. A sinogram for displaying SPECT data can be seen in Figure 7-11. In the previous example we saw that the sinogram was a different way to look at acquired projection views. The sinogram is used in a very similar manner

to display PET data. A simplified illustration of the layout of the matrix of detector pairs used to create a PET sinogram for an imaginary camera with 12 detectors is seen in Figure 10-18. Each row of the sinogram contains the counts for all of the opposing detector pairs that have parallel lines of response (the paths between the detectors).

The matrix elements associated with a single detector in this simple system are highlighted in grey and create a diagonal line in Figure 10-19. A normal blank sinogram from a ring detector is shown in Figure 10-20A. A defective detector appears as a new thin diagonal line across the sinogram (illustrated in Fig. 10-20B).

PMT Gain Test

The light output from each crystal is predictable when a known quantity of a positron emitter

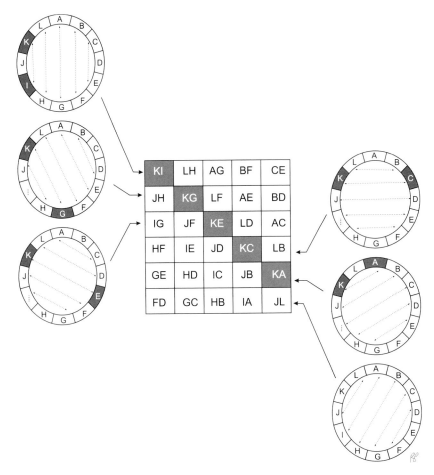

Figure 10-19 Individual detector in sinogram matrix.

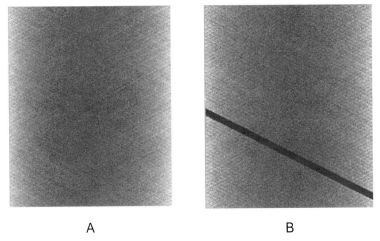

Figure 10-20 (A) Normal blank transmission sinogram. (B) Blank transmission sinogram with a defective detector (Courtesy of Frederick Fahey, D.Sc.).

such as ^{68}Ge or ^{22}Na positron source is centered within the gantry opening. The magnification of a signal from a few electrons to millions of electrons, or the signal gain that occurs within a photomultiplier tube when light photons strike the photocathode end, should also be reproducible (see Fig. 5-5 and associated text). However, largely due to temperature and humidity changes, the PMT response can vary over time. The output from each PMT is checked *daily* or *weekly* to assure that the response across the PMTs is uniform.

Normalization

This *quarterly* procedure is used to adjust the PMT gains. Since most PET systems have hundreds or thousands of individual crystals a long acquisition is needed to collect adequate statistics to measure a precise gain for each PMT. An overnight (10-15 h) transmission scan with an empty gantry is obtained. All of the PMTs gains are then automatically internally adjusted using these data.

Calibration

This *quarterly* procedure is used to calibrate the system response to a known amount of radioactivity in a known volume. In other words, a certain number of PET counts within a pixel can be converted to the average activity concentration (in kBq/ml) for that pixel. Calibration is necessary for quantitative measurements such as standard uptake values (see Chapter 8). A small amount of measured positron emitter is uniformly distributed in a phantom of known volume. The phantom is imaged. The counts from the attenuation corrected images of the phantom can then be correlated to the activity in mBq/ml.

Reference

1 NUREG-1556, Program Specific Guidance About Medical Use Licenses. Final Report, Vol. 9, October, 2002. Division of Industrial and Medical Nuclear Safety, Office of Nuclear Material Safety and Safeguards, U.S. Nuclear Regulatory Commission, Washington, D.C. Appendix K.

Test Yourself

1 Match each of the following quality control procedures with the recommended frequencies of performance:
  (a) Dose calibrator linearity
  (b) Dose calibrator constancy
  (c) Survey meter constancy
  (d) Thyroid probe efficiency
  (e) Well counter calibration.
  Choices:
   (i) daily
   (ii) at installation, quarterly, and after repairs
   (iii) at installation, annually, and after repairs.

2 True or false: A 5-million count daily uniformity flood can be used to create the uniformity correction matrix for SPECT imaging.

3 Which of the listed QC procedures might detect the following camera problems (more than one answer per question is possible)?
  Camera problem:
  (a) drift of the energy window
  (b) malfunction of a photomultiplier tube
  (c) collimator damage
  (d) decrease in spatial resolution.
  Quality control procedure:
   (i) extrinsic flood
   (ii) intrinsic flood
   (iii) flood source with bar phantom
   (iv) checking the photopeak.

4 Associate each of the following camera correction factors with the most applicable procedure in the lettered list:
  Camera correction factor:
  (a) PET uniformity correction
  (b) attenuation correction

(c) SPECT camera resolution

(d) center of rotation.

Procedure:

(i) empty port scan

(ii) transmission scan

(iii) emission scan

(iv) sinogram

(v) tomographic scan of phantom

(vi) projection imaging of point source.

5 True or false: In assessing the results of the center of rotation test, a deviation of the measured position of the source of less than two pixels from the expected position does not need further investigation.

CHAPTER 11

11 Radiation biology

The biologic effect of radiation can be understood in terms of the transfer of energy from the radiation (photons and particles) to the tissue. When the energy of radiation is deposited in the body, it can disrupt the chemical bonds and alter tissue. It is important to understand some of the details of this transfer. The interaction of radiation and tissue is governed by the energy and mass of the incident radiation (alpha and beta particles, gamma ray, or x-ray) and the properties of the tissue.

Radiation Units

Radiation Absorbed Dose (rad)

The radiation absorbed dose, or **rad**, is a measure of energy transferred to any material from ionizing radiation. The corresponding système international (SI) unit is the **gray(Gy)**, named after the English physicist Louis Harold Gray. Remember that ionizing radiation is a term that applies to any radiation that is sufficiently energetic to create ion pairs; it includes x-rays, gamma rays, alpha and beta particles, and neutrons and protons.

One rad is equal to 100 ergs of energy absorbed per gram of tissue; 1 gray is equal to 100 rads. For example, a bladder wall receives approximately 0.04 Gy (4 rad) from its contained urine following the excretion of a 740 MBq (20 mCi) intravenous dose of ^{99m}Tc-pertechnetate.

Table 11-1 Quality Factors (QF) for Ionizing Radiation

Ionizing Radiation	QF
Alpha	20
Protons, neutrons	10
Beta (electrons and positrons)	01
Gamma and x-rays	01

Roentgen-Equivalent-Man (rem)

The roentgen-equivalent-man, or **rem**, is the unit of absorbed energy that takes into account the estimated biologic effect of the type of radiation that imparts the energy to the tissue. Particles with higher linear energy transfer, such as alpha particles, protons and neutrons, produce greater tissue damage per rad than beta particles, gamma rays, or x-rays. The relative damage for each type of radiation is referred to as its **quality factor (QF)**, the values for which are given in Table 11-1.

The **dose equivalent** (or **absorbed energy**) is the product of the dose times the quality factor, or

Dose equivalent (in rem)

$$= \text{absorbed dose (in rad)} \times \text{QF}$$

And, in SI units,

Dose equivalent (in sievert)

$$= \text{absorbed dose (in Gy)} \times \text{QF}$$

Table 11-2 Quantities and Units Used in Nuclear Medicine

Quantity	Système International (SI) Unit	Conventional Unit	Equivalence	Meaning
Activity	becquerel (Bq)	curie (Ci)	$1 \text{ Bq} = 2.7 \times 10^{-11} \text{ Ci}$	Number of disintegrations of radioactive material per second
Absorbed dose ionizing	gray (Gy)	rad	$1 \text{ Gy} = 100 \text{ rad}$	Energy absorbed from radiation per unit mass of absorber
Exposure by	coulombs per kilogram (C/kg)	roentgen (R)	$1 \text{ C/kg} = 3.9 \times 10^3 \text{ R}$	Amount of charge liberated ionizing radiation per unit mass of air
Dose equivalent	sievert (Sv)	rem	$1 \text{ Sv} = 100 \text{ rem}$	Absorbed dose times the quality factor (Dose × QF)

The quantities and units used to measure radiation in nuclear medicine are given in Table 11-2.

UNITS OF ENERGY AND FORCE

One erg (from the Greek word for work) is the amount of energy imparted to a body by a force of 1 dyne acting over a distance of 1 cm. The dyne (also from the Greek and with the same root as "dynamo") is simply the amount of force required to give a mass of 1 g an acceleration of 1 cm/s^2.

One erg is only a small amount of energy in everyday terms, and, as you might reasonably conclude, the dyne is a similarly small amount of force. Both are units of the centimeter-gram-second (CGS) system.

The system of units whose magnitude is more familiar is the meter-kilogram-second (MKS) system. The MKS units of energy and force are the joule ($=10^7$ ergs) and the newton ($=10^5$ dynes). The units of the SI system are the same as the MKS system.

In terms of more familiar energy units, 1 J equals 0.24 small calories (as opposed to the kcal used to measure food consumption). One erg is only one ten-millionth of a joule.

The Effects of Radiation on Living Organisms

The effects of radiation on living organisms can be described at the level of the cell, a tissue, an entire organism, or a whole population.

Cellular Effects

Individual Cells
Cellular Structure
Cells are the building blocks for living matter and are composed of a nucleus and cytoplasm. The nucleus contains the genetically important chromosomes that are composed of **deoxyribonucleic acid (DNA)**, which is a large molecule consisting of thousands of small subunits (bases) coiled into a double helix. The genetic code for the cell and the entire organism is held in the sequence of pairs of bases along the two strands of the double helix. During the process of cell division, called **mitosis**, the DNA reproduces itself so that a complete set of chromosomes is deposited within each cell. In this way the genetic code is propagated. Since DNA plays such a pivotal role

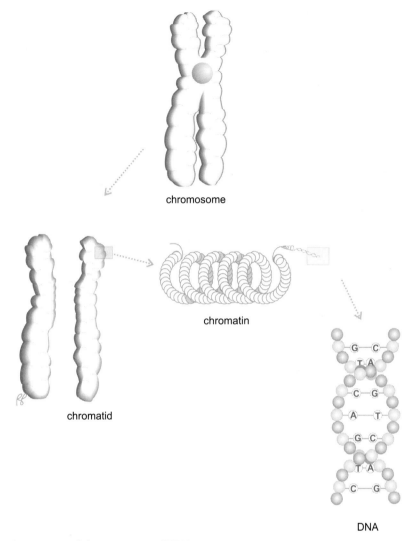

Figure 11-1 The structure of chromosomes and DNA.

in cellular multiplication and function, radiation damage to DNA has a profound impact on living tissue. Chromosomes are particularly radiosensitive (vulnerable to radiation damage) during mitosis.

DNA contains two chains of alternating sugar molecules and phosphate groups. The sugars of the chains are linked by pairs of **nucleotides**—the "bases" thymine, cytosine, adenine, or guanine. The term base refers to the basic, as opposed to acidic, nature of the isolated compounds. These four bases are commonly referred to by their initial letters—T, C, A, and G. The chains and nucleotides are arranged in two long, coiled, and intertwined strands, which are tightly packed to form a **chromatid** (Fig. 11-1). Two identical chromatids are "attached" to a centromere and form a **chromosome**. As shown at the bottom of Figure 11-1, the bases form pairs across the strands as follows: adenine pairs with thymine and guanine with cytosine. The genetic code for the cell and, indeed, for the entire organism is held in the sequence of the nucleotides with their base pairs.

FREE RADICALS

Indirect DNA damage is mediated by free radicals. When particulate or photon radiation interacts with water (H_2O) an ion pair is formed (H_2O^+, e^-). The electron will combine with H_2O to form H_2O^-. The H_2O^+ and H_2O^- are called ion radicals (not to be confused with free radicals). Ion radicals are very unstable and rapidly dissociate: H_2O^+ becomes H^+ and $OH^\bullet$, and H_2O^- becomes $H^\bullet$ and OH^-. $OH^\bullet$ and $H^\bullet$ are free radicals.

A free radical is an atom or molecule that has no electrical charge but is highly reactive because it has an odd number of electrons with an unpaired electron in its outer shell. Free radicals tend to quickly recombine to form stable electron configurations. However, in high enough concentrations in the cell, they can create organic free radicals ($R^\bullet$) and $H_2O_2^\bullet$ (hydrogen peroxide), a toxic molecule. $OH^\bullet + RH$ become $R^\bullet + H_2O$ and two $OH^\bullet$ become H_2O_2. Organic free radicals in DNA lead to breakage of the strands and cross-linking. $OH^\bullet$, since it oxidizes (removes electrons), is more damaging than $H^\bullet$, which is a reducing agent (gives up its electrons).

Mechanisms of Radiation Damage to DNA

Ionizing radiation can cause deletions or substitutions of bases and/or actual breaks in the DNA chain. Ionizing particles (neutrons, alpha and beta particles) are responsible for biologic damage. Photons (x-rays and gamma rays) transfer energy to "fast" electrons (via Compton scattering and the photoelectric effect), which in turn cause biologic damage through ionization.

Base deletions or substitutions can have variable effects on the cell line. DNA strand breaks, if not repaired, cause abnormalities in chromosomes that may result in cell death. **Single breaks**, caused by low LET (linear energy transfer; see Chapter 3) radiation given at a low dose rate, are relatively easily repaired by using the other strand of DNA as a template. Radiation of relatively high LET, or a high dose rate of low LET, may produce single breaks in close proximity to each other in both strands (called **double or multiple strand breaks**), which are more difficult to repair (Fig. 11-2).

Direct and Indirect Action of Radiation

DNA damage can occur as a result of a **direct** action in which the particulate radiation (such as alpha particles) or fast electrons produced by photons in Compton and photoelectron interactions strike the DNA molecules (Fig. 11-3). Alternatively, damage may be caused by **indirect** action in which the radiation interacts with the water molecules in the cell to form free radicals, which in turn damage the DNA strands (see Fig 11-3). Most DNA damage caused by low LET radiation is a result of indirect action. Most DNA damage by high LET radiation is via direct action.

Radiosensitivity and Cell Cycle

The normal mitotic cycle of the cell is illustrated in Figure 11-4. The **cell cycle** can be divided into four segments, the length of which varies as a function of cell type. During **mitosis** the cell divides into two individual cells. This is followed by **interphase G_1** during which only one copy of the cellular DNA is contained in a chromatid. The **synthetic (S) phase** is a period of DNA replication. In **interphase G_2** the chromatid has duplicated and the DNA is doubled and is located in chromosomes.

Experiments have shown that the cell is relatively resistant to radiation damage during the latter part of the S phase, the period of DNA synthesis. It is hypothesized that during this portion of the cell cycle, the relative abundance of repair enzymes (DNA polymerase and ligase) facilitates DNA repair. Other experiments have shown that the greatest amount of damage occurs during the period of mitosis; the dose needed to halt mitotic activity in a dividing cell is much less than that needed to destroy the function of a differentiated cell. The latter portion of the G_2 phase is nearly as sensitive as the mitotic phase.

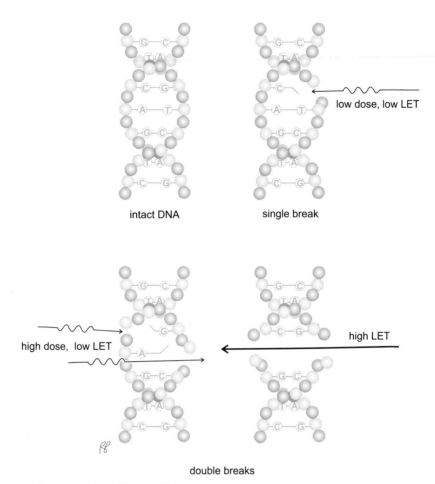

intact DNA

single break

low dose, low LET

high dose, low LET

high LET

double breaks

Figure 11-2 Single strand and double strand breaks.

Cell Population Effects

Cell Survival Curves

Thus far we have outlined the effects of radiation on individual cells. Since it is likely that cells will be randomly damaged, it is more useful to consider the overall effects of radiation on groups of cells.

The effects of radiation on cell populations are expressed as a plot of the fraction of surviving cells vs. radiation dose to the population of cells. Cell survival as a function of radiation dose is linear for high LET radiation; this is because most DNA damage is via multiple breaks that are generally not repaired. Cells exposed to lower doses of low LET radiation can often repair the single strand breaks caused by this radiation.

For the lower doses of this radiation, therefore, the fraction of the cells surviving is greater than for a similar dose of high LET radiation. The initial shoulder in the low LET curve reflects the cells' ability to repair the single breaks caused during low doses of low LET radiation (Fig. 11-5).

Conventional terms used to describe these curves are given in Table 11-3.

Factors Affecting Cell Survival

Dose Rate

Low LET radiation: The degree of damage incurred by administering a dose of low LET radiation is dependent on the rate at which this dose is delivered. With delivery of a given dose over a longer period of time (lower dose rate), most

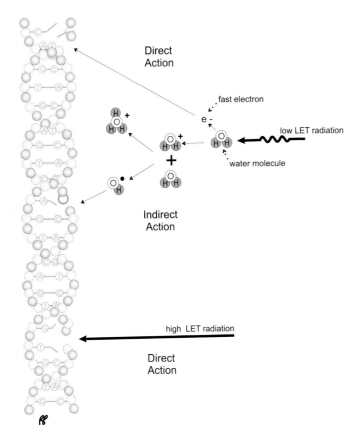

Direct
Action

fast electron

e -

low LET radiation

water molecule

+

Indirect
Action

high LET radiation

Direct
Action

Figure 11-3 Direct and indirect action.

DNA damage is via single breaks and the cell has time to repair the damage. At high dose rates there are more double breaks and less time to repair single breaks. As a result, the shoulder of the curve (Dq) approaches 0, and the extrapolation value (n) approaches 1 (Fig. 11-6).

High LET radiation: High LET radiation causes such a high incidence of double breaks that repair is negligible at any rate.

Chemical Interventions
The introduction of certain chemicals in the medium in which the cells exist can alter the cell population's response to administered radiation doses. These chemicals alter the indirect effects (via free radicals) of radiation on DNA.

Radiosensitizers: These substances increase the amount of radiation damage to cells at a given radiation dose and are used in radiation therapy to increase death of tumor cells. Radiosensitizers include oxygen, halogenated pyrimidines, and nitroimidazoles.

Oxygen enhances the indirect action of local LET radiation by binding to the free radical $R^\bullet$ on the damaged ends of a DNA break. The resulting $RO_2^\bullet$ is a more stable free radical, and the damaged ends are less likely to be repaired (Fig. 11-7). Oxygen has less effect on the radiotoxicity of high LET radiation, which is more likely to damage DNA by direct action. The effects of oxygen as a radiosensitizer is most pronounced in anaerobic tissues such as the center of tumor masses.

The **oxygen enhancement ratio (OER)** is the ratio of the dose in hypoxic tissue to the dose in aerated tissue required to cause a given tissue effect. For photon or x-ray irradiation, this ratio is approximately 2.0 to 3.0. For high LET radiation (alpha particles), the ratio approaches 1.0, since most of the damage is via direct action.

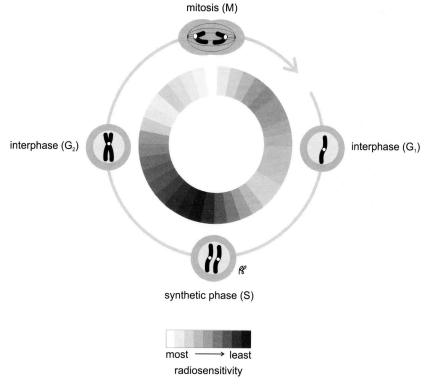

Figure 11-4 The cell cycle and relative radiosensitivity.

Figure 11-8 illustrates an OER of 2.0 from a hypothetical x-ray dose. Half of the x-ray dose is needed during oxygenation to kill the same proportion of the cell population radiated in anaerobic conditions.

The **halogenated pyrimidines**, 5-iododeoxyuridine and 5-bromodeoxyuridine, are very similar to thymine and are easily incorporated in its place into DNA. However, the substitution "weakens" the DNA making it more susceptible to radiation damage.

The **nitroimidazoles**, including misonidazole, and etanidazole, mimic the action of oxygen by binding to the free radical R$^\bullet$ on the damaged ends of a DNA break. They are effective in hypoxic tumors as they can penetrate more deeply; unlike oxygen they are not metabolized by surrounding tissue.

Radioprotectors: **Radioprotectors** such as cysteine and amifostine (also known as WR-2721)

"scavenge" or combine with free radicals, thereby reducing the likelihood of low LET DNA damage.

Tissue Effects

The radiosensitivity of different types of tissues is largely dependent on the mitotic rate of the cells comprising the tissue and their degree of differentiation. In general, cells with a higher mitotic rate and those that are less differentiated are more radiosensitive. For poorly understood reasons, mature lymphocytes are an exception to the general rule and are very radiosensitive. (It has been hypothesized this is because their large nucleus makes the DNA an easy target.) Table 11-4 lists cell types and their relative radiosensitivity.

Organ Toxicity

Table 11-5 outlines the effects of acute exposure on selected organs.

Embryo and Fetus

The developing human in utero is most likely to suffer severe and life-threatening effects of even low doses of radiation during the period of major

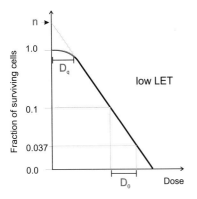

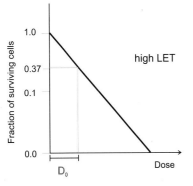

Figure 11-5 Cell survival curves for low and high LET radiation.

organogenesis (weeks 3 to 6). Prior to day 11, when the cells are pluripotential, the preimplantation embryo will either succumb to radiation or will survive without significant abnormalities. After week 6, the fetus is relatively radioresistant; however, the radiosensitive neuroblasts and germ cells may be damaged (Casarett 1968, pp. 231–232). Table 11-6 outlines the effects of radiation during these phases of development.

Therapeutic abortion should be considered when the fetus receives a dose of 10 rads (0.1 Gy) or more between day 11 and week 26 postconception.

Acute Whole-Body Radiation Toxicity

Radiosensitivity varies between species. It can be expressed as the **50% lethal dose**, or **$LD_{50/30}$**, which refers to the radiation dose that kills 50% of the population within 30 days. For humans, the $LD_{50/30}$ is 2.5 to 4.5 Gy (250 to 450 rad); for mice, it is 5.5 Gy (550 rad) (Casarett 1968, p. 220).

Acute radiation sickness following whole body radiation is discussed in Chapter 14.

Cancer and Genetic Effects

The adverse effects of radiation exposure include death and tissue damage (as described above), the development of cancer, and inherited genetic effects. The risk of developing an adverse effect is classified as either stochastic or nonstochastic.

Table 11-3 Terms and Definitions for Cell Survival Curves

Term	Definition	Significance
D_0	The dose that decreases the population by 63%, this term is used for high LET curves and the linear portions of low LET curves	More radiosensitive cells have a lower D_0
D_q	The quasi-threshold is the width of the "shoulder" or initial nonlinear portion of the low LET survival curves	Cells with greater ability to repair single breaks of low-dose LET are more radioresistant and have a larger D_q (for high LET, D_q approaches 0)
n	The extrapolation number is the relative "survival fraction" if the linear portion of the survival curve is projected backward toward the y-axis	The higher the extrapolation number the steeper the dose response curve, and the greater the radiosensitivity of the cells to low LET radiation. High LET radiation have an n of 1.0.

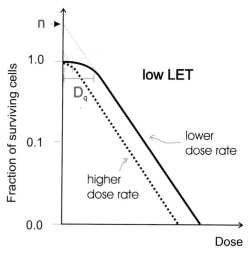

Figure 11-6 Effect of dose rate on cell survival.

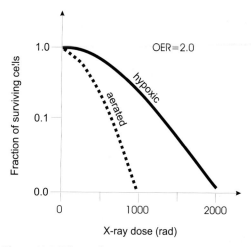

Figure 11-8 Effects of oxygenation on cell survival during low LET radiation.

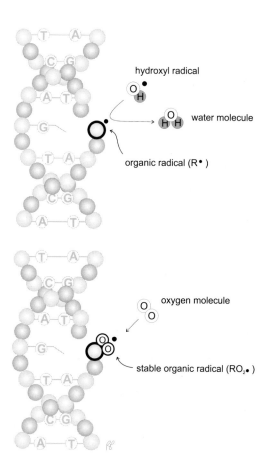

Figure 11-7 Oxygen binds the organic radical created during interaction with a hydroxyl radical.

Table 11-4 Radiosensitivity of Cell Types

Relative Radiosensitivity	Examples
Highly sensitive	Spermatogonia, erythroblasts, lymphocytes
Relatively sensitive	Epidermal basal cells, intestinal crypt cells, myelocytes
Intermediate sensitivity	Osteoblasts, spermatocytes, chondroblasts
Relatively resistant	Liver cells, spermatozoa, granulocytes, erythrocytes, osteocytes
Highly resistant	Neural cells, muscle cells

Stochastic and Nonstochastic Risks

Cancer and genetic effects of radiation are examples of **stochastic risks**. The likelihood that an individual will develop cancer or suffer from her parents gonadal exposure to radiation is dose dependent; the greater the dose the more likely the adverse effect. However, these are only statistical probabilities; it is not possible to know that the individual will develop cancer or a genetic defect (unless of course the dose is high enough to create a 100% risk). For example, it can be predicted that 1 out of every 100 children

Table 11-5 Organ Toxicity from Acute Exposure

Organ	Type of Damage	Threshold Gy (rad)
Ocular lens	Cataract formation	2 (200)[a]
Skin	Severe damage with intense erythema, wet desquamation, permanent hair loss, loss of sweat glands, vascular damage, ulceration	20–50 (2000–5000)
Reproductive Organs	Permanent male sterility	5 (500)
	Permanent female sterility	6 (600)

Note:

[a] There is a latency period for cataract formation of up to 8 years for doses up to 6.5 Sv (650 rad) (Hall 1994, p. 383).

Table 11-6 Radiosensitivity of the Embryo and Fetus

Stage	Characteristics	Time Period	Effect
Early embryo	Preimplantation	Days 0–11	"All or nothing": termination or survival without apparent defects
Late embryo	Major organogenesis	Weeks 3–6	Potentially severe organ abnormalities including growth and mental retardation
Fetus	Minor organogenesis	After week 6	Less severe organ abnormalities, such as eye, skeletal, and genital defects; possible microcephaly, as well as growth and mental retardation

born to parents who were exposed to a certain amount of radiation will develop leukemia; it is not known which of these children will develop leukemia. Child X has a 1/100 risk of developing leukemia, but it is not known whether she will develop leukemia over her lifetime.

Other risks are directly dose-related. These are referred to as **nonstochastic** (also called **deterministic**) **risks**. For nonstochastic risks there are thresholds above which radiation damage can be expected in all exposed individuals. For example, if one of the parents received 6 Gy (600 rad) to his eyes, it is known that as a result of this radiation he will develop cataracts in the future. The risk is said to be 100% above a threshold of 2 Gy (200 rad).

Genetic Effects

Gonadal Dose

The **gonadal dose** is the calculated dose to the gonads for a given individual and is a sum of doses derived from natural and man-made radiation exposures. All individuals in the population receive about 1 mSv/year (100 mrem/year) of background radiation to their gonads. A radiation worker might receive an additional gonadal dose of approximately 5 mSv/year (500 mrem/year). His total gonadal dose would be approximately 6 mSv/year (600 mrem/year). A patient with a chronic illness might receive an average of 1 mSv/year (100 mrem/year) of exposure from x-rays; her gonadal dose would then be approximately 2 mSv/year (200 mrem/year).

Genetically Significant Dose

In order to determine the relative contributions of all of the above sources of radiation to the incidence of genetic birth defects, a value called the **genetically significant dose (GSD)** has been developed. The GSD for background radiation is the same for all individuals and equals the gonadal dose, which is 1 mSv/year (100 mrem/year).

To calculate the GSD from man-made radiation, the total gonadal dose for the subset of the population receiving occupational and medical exposure can be estimated. If this total dose is then distributed, or averaged, over the entire population (with significantly greater weight being given to doses received by the fertile portion of the population), a dose of approximately 0.25 mSv/person-year (25 mrem/person-year) is calculated. The total GSD is then a sum of the background dose and the man-made dose or 1.25 mSv/year (125 mrem/year).

The baseline genetic defect rate is 1000 severe defects per 1 million live births. The estimated excess severe genetic defects from radiation (calculated by the Committee on the Biologic Effects of Ionizing Radiation (BEIR) are 100 occurrences per 1 million live births per rem. A GSD of 0.5 to 2.5 Sv/year (50 to 250 rem/year) is necessary to double the baseline genetic defect rate; this value is referred to as the **doubling dose**.

Carcinogenic Effects

The incidence of certain cancers—such as leukemia, head, neck, pharyngeal, thyroid, breast, and lung—historically has been shown to be increased following radiation exposure (as observed in populations following radiation disasters such as the bombing of Hiroshima). The risk to an individual whose exposure history is known can be estimated by multiplying the absorbed dose by the **risk factor**. The risk factor currently used by the U.S. Nuclear Regulatory Commission (NRC) for cancer induction for the adult working population exposed to low dose, low dose rate radiation is 4×10^{-2} cases of cancer per sievert (4×10^{-4} cases per rem). The rate for the general population, including the young, is somewhat higher at 5×10^{-2} cases of cancer per sievert [1].

To illustrate the application of the risk factor, consider a medical technologist who has received 2.5 mSv/year (250 mrem/year) for 20 years. The lifetime dose to this individual is 20 years × 0.25 rem = 5 rem. The risk of cancer is 4×10^{-4} cases/rem × 5 rem = 0.002. Thus, there is a 0.002 (0.2%) increase of cancer risk for this individual. The natural incidence of fatal cancer is about 20%, so this exposure to radiation has increased the risk of fatal cancer for this individual from 20% to 20.2%. To put this calculation in perspective, however, the risk of cancer associated with 30 years of exposure at the maximum permissible total effective dose equivalent is lower than the risk of a fatal automobile accident but higher than the risk of a fatal electric shock.

Reference

1 Hall, EJ, *Radiobiology for the Radiologist*, 5th Edition, Lippincott Williams and Wilkins, Philadelphia, 2000, p. 157.

Test Yourself

1 Which of the following statements are correct for radionuclides used in nuclear medicine imaging:

(a) The dose equivalent is ten times the absorbed dose

(b) The dose equivalent is equal to the absorbed dose

(c) The quality factor, or relative tissue damage per administered Gy of radiation, is larger for high LET radiation than low LET radiation

(d) All of the above.

2 True of false: Single strand DNA breaks which are caused by low dose low LET radiation are less likely to be repaired by a cell than double or multiple strand breaks which are caused by high dose low LET radiation or high LET radiation.

3 True or false: The shoulder seen in the cell survival curves of low dose low LET radiation has been attributed to the ability of the cell to "keep-up" with the repair of single strand DNA breaks.

4 Rank the following phases of the cell cycle from the most radiosensitive to the least radiosensitive:
   (a) Mitosis
   (b) Interphase G1
   (c) Synthetic phase, S
   (d) Interphase G2.

5 Connect the following doses or risk factors with their estimated values:
   (a) Threshold dose in Gy above there is 100% incidence of cataracts
   (b) Genetically significant dose (GSD) that will increase the genetic defect rate to twice the baseline rate (called the doubling dose)
   (c) Risk factor for cancer induction in the adult working population per sievert of low dose, low dose rate radiation exposure.

Dose or risk factor value:
   (i) 0.5 to 2.5 Sv/year (50 to 250 rem/year)
   (ii) 4×10^{-2} (4×10^{-4} cases per rem)
   (iii) 2 Gy (200 rads).

12 Radiation dosimetry

Dosimetry is the calculation of the total absorbed dose to individual organs or the whole body from internal radiation exposure. Dosimetry information is supplied with all radiopharmaceuticals, as shown for thallium-chloride in Table 12-1.

Physical, Biologic, and Effective Half-Lives

The calculation of dosimetry relies on an understanding of the different types of half-lives used to describe radiopharmaceuticals. The **physical half-life** (T_p or $T_{1/2}$) is the time it takes for half of the nuclide atoms to become stable (see Chapter 1). The **biologic half-life** (T_b) has nothing to do with radioactivity, but rather reflects the half-time for excretion of the material from the organ or whole body. For instance, the biologic half-life of ^{99m}Tc-MDP is the time it takes for one half of this radiopharmaceutical to be filtered and excreted by the kidneys and bladder. The **effective half-life** (T_e) is a measurement that combines the above two values; it is the time required for one half of the initial radioactivity to disappear from an organ or the body both by excretion and physical decay. The effective half-life is always shorter than either the physical or biologic half-life and is calculated using the formulas

$$1/T_e = 1/T_b + 1/T_p \qquad (Eq.\ 12\text{-}1)$$

Table 12-1 Part of the Estimated Dosimetry for a 74 MBq (2-mCi) Dose of ^{201}Thallium-chloride

Tissue	mGy/74 MBq	rad/2 mCi
Kidneys[b]	34.0	3.40
Thyroid[b]	46.0	4.60
Liver[b]	7.4	0.74
Testes[b]	60.0	6.00
Total body[a,c]	4.2	0.42
Effective Dose Equivalent[b,c]	27 mSv	2.6 rem

Note:
[a] Package insert from Thallous Chloride Tl 201, E.I. duPont de Nemours, 1988.
[b] Package insert from Thallous Chloride Tl 201, Mallinckrodt, 2004.
[c] **Total body dose** is defined as the total energy deposited in the body divided by the mass of the body, the **effective dose equivalent** is the sum of the weighted individual organ doses.

Or

$$T_e = (T_b \times T_p)/(T_b + T_p) \qquad (Eq.\ 12\text{-}2)$$

Table 12-2 lists hypothetical values to demonstrate the relationship between the three types of half-lives. Table 12-3 lists actual values for selected radiopharmaceuticals.

Table 12-2 Physical, Biologic, and Effective Half-Lives

Tp	Tb	Te
1000	1	00.999
200	100	67.000
10	10	05.000
1	1	00.500
10	20	06.700
1	1000	00.999

Table 12-3 Sample Physical, Biologic, and Effective Half-Lives

Radiopharmaceutical	Tp	Tb	Te
^{99m}Tc-sulfur colloid	6 h	∞^a	6 h
^{99m}Tc-MDP	6 h	4 h	2 h
^{67}Ga-citrate	78 h	530 h^b	68 h
^{123}I	13 h	26 h	8.7 h
^{131}I (30% uptake)	8 days	70 days	7 days

Note:

^{a} Sulfur colloid is taken up by reticuloendothelial cells and has a very slow elimination from the liver, so the biologic half-life of the sulfur colloid is estimated as infinite.

^{b} A weighted average of fast and slow components.

Source: Data from Loevinger, R, Budinger, TE, and Watson, EE, *MIRD Primer for Absorbed Dose Calculations*. Revised edition, Society of Nuclear Medicine, New York, 1991, p. 53; Gollnick, DA, *Basic Radiation Protection Technology*. 2nd edition, Pacific Radiation, Altadena, CA, 1988; and Loevinger, R, Budinger TE, Watson EE. *MIRD Primer for Absorbed Dose Calculations*. Revised edition, Society of Nuclear Medicine, New York, 1991, p. 40.

Calculation of Dosimetry

The dose delivered to an organ, $\overline{D}$, is equal to the total (cumulated) activity within the organ, $\tilde{A}$, times **S**, a correction factor based on characteristics of the administered radionuclide

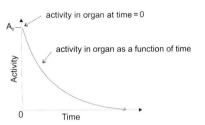

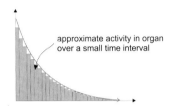

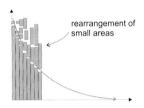

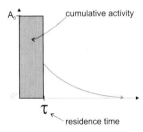

Figure 12-1 Time-activity curve, cumulated activity, and residence time.

and the organ of interest.

$$\overline{D} = \tilde{A} \times S \qquad (Eq.\ 12\text{-}3)$$

Description of Terms

A_0, the **initial activity**, is the amount of injected activity that is in the organ immediately after injection. **A(t)** is the activity within the organ as a function of time. The **cumulated activity**, $\tilde{A}$, is the summed total activity within the organ over the entire time the radioactivity remains within the organ. The uppermost plot in Figure 12-1 is

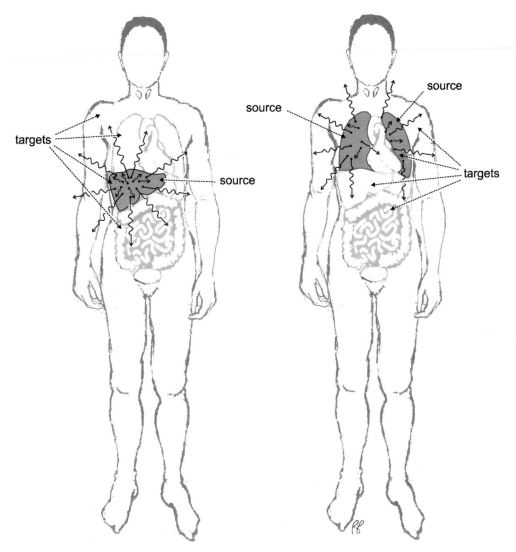

Figure 12-2 Source and target organs.

a time activity curve; the activity at time 0 is A_0. The area under the curve, $\tilde{A}$, can be approximated with sequential thin rectangles (second uppermost panel). The cumulated activity can be calculated from the initial activity and the effective half-life, T_e, of the radiopharmaceutical as follows.

The residence time, τ, is the time over which the organ would receive the same total dose from the amount of initial activity (A_0) were this to remain constant, and then fall instantly to zero. If one were to rearrange the thin rectangles in the second panel of Figure 12-1 into a large rectangle of height A_0, the width of this rectangle would be τ (lower panels).

$$A = \tau \times A_0 \qquad (Eq.\ 12\text{-}4)$$

The residence time is related to the effective half-life (discussed above) by the equation

$$\tau = 1.44 \times T_e \qquad (Eq.\ 12\text{-}5)$$

It follows then that

$$\tilde{A} = 1.44 \times T_e \times A_0 \qquad (Eq.\ 12\text{-}6)$$

S Value

The radiation dose, $\overline{D}$, to any organ depends on the amount of activity cumulated in that organ ($\tilde{A}$), on the size, shape, and density of the organ, and on the energy and type of radiation it contains. We have already discussed how cumulated activity can be calculated, but the calculation for the remaining factors is too complicated to pursue here. Fortunately these calculations have been performed for most of the organs of the body and for the range of photon and particle energies emitted by the medically important radionuclides. These combined factors are referred to as the S value, which is available in standard tables.

An in-depth demonstration of the calculation of an S factor is presented in Appendix C.

Self-Dose, Target, and Source Organs

The term **self-dose** is used to describe the radiation dose to an organ from the radiation within it, for example, the dose to the liver from the accumulation of ^{99m}Tc sulfur colloid within the liver. Moreover, the same approach can be applied for calculating the dose to any other organ of the body, the **target**, from the radioactivity of any other organ, called the **source** (for example, the dose to the thyroid from ^{99m}Tc sulfur colloid in the liver). The left side of Figure 12-2 depicts the liver as the source organ and all organs, including the liver, as target organs. In the figure on the right, the lung is the source organ following injection of ^{99m}Tc-MAA. The total dose to any organ as target is the sum of doses from all sources within the body.

SAMPLE CALCULATION OF $\overline{D}$

A sample calculation of an absorbed dose may elucidate the above concepts. We will use a hypothetical case of ^{131}I sodium iodide ingestion with the lung as the target organ and the thyroid as the source organ. To simplify the calculation, we will assume that the thyroidal uptake of ^{131}I sodium iodide is instantaneous and that the 24-hour radioiodine uptake is 30%. The ingested dose is 37 MBq (1 mCi).

The initial activity in the thyroid is

$$A_0 = 37 \text{ MBq} \times 30\% \text{ uptake}$$
$$= 12.3 \text{ MBq} (0.3 \text{ mCi})$$

The effective half-life (see Table 12-2) is

$$T_e = 7 \text{ days}$$

The residence time is

$$\tau = 1.44 \times T_e$$
$$= 10 \text{ days}$$

The cumulated activity is

$$A = \tau \times A_0$$
$$= 10 \text{ days} \times 37 \text{ MBq}$$
$$= 370 \text{ MBq-days}$$
$$= 2952 \text{ MBq-hr} = 72 \text{ mCi} - \text{hr}$$

The S value (MIRD pamphlet No. 11) for source thyroid and target lung is calculated as $2.9 \times 10^{-6} \text{ rad/uCi/hr}$. The absorbed dose to the lungs from the thyroid is

$$\overline{D} = \tilde{A} \times S$$
$$= 72,000 \text{ uCi-hr} \times 2.9 \times 10^{-6} \text{ rad/uCi/hr}$$
$$= 0.21 \text{ rad} (0.0021 \text{ Gy})$$

Test Yourself

1 True or false: The effective half-life is always longer than either the physical or biological half life.

2 What is the effective half-life of rubidium-86 which has a biological half-life of 45 days and a physical half-life of 18.8 days?

3 True or false: The biological half-life of a radiopharmaceutical labeled with ^{99m}Tc will be shorter than the same compound labeled with ^{111}In.

4 True or false: When calculating the total dose to a target organ the target organ should be included as one of the radiation sources.

13 Radiation safety

Rationale

The purpose of a radiation protection program is to monitor individuals' contact with radiation and to limit their exposure to as low a level as possible. Federal regulations are issued by the Nuclear Regulatory Commission (NRC) that outline acceptable levels of exposure. In addition, the government has set forth a general policy principle referred to as **ALARA**. To quote from 56 Federal Register No. 98:

> ALARA (acronym for "as low as reasonably achievable") means making every reasonable effort to maintain exposures to radiation as far below the dose limits . . . as is practical consistent with the purpose for which the licensed activity is undertaken.

Dose Limits

Radiation dose may come from radiation sources outside the body (for example, an x-ray machine or radioactive material external to the body) or from radioactive material that has been taken into the body. These two modes of exposure to radiation are called **external exposure** and **internal exposure**, respectively. Table 13-1 lists the terms commonly used to describe radiation exposure.

Dose limits are separately prescribed for occupational workers (including those in the nuclear medicine department), the general public, and the fetus.

Occupational Exposure

The prescribed limits for **occupational exposure for radiation workers** are listed in Table 13-2.

Hospital Workers

The maximal permissible exposure to **hospital workers who are not classified as radiation workers** is the same as the general public: 1 mSv/year (0.1 rem/year). If an individual is likely to receive over 10% of any of the acceptable limits, they are classified as radiation workers and must be monitored (for example, by using a pen dosimeter or wearing a film and/or ring badge). For those individuals receiving less than this amount, their exposure is estimated.

Exposure to the General Public

The following rules apply to **public exposure** from radioactive materials (including patients containing radioactive materials). The estimated **annual total effective dose equivalent (TEDE)** for a member of the public should be less than 1 mSv (0.1 rem). The estimated

Table 13-1 Terms Used To Describe Exposure[a]

Term	Abbreviation	Description
Absorbed dose	D	Dose of nuclide (MBq or rad)
Quality factor	QF	Relative biologic effectiveness of type of radiation; QF = 1 for gamma and beta emissions[b]
Dose equivalent	H_T (DE)	The whole body or organ dose (mSv or rem); $H_T = D \times QF\ (\times N)$[c]; for nuclear medicine exposures $H_T = D$
Deep dose equivalent	H_d (DDE)	External exposure to the whole body as measured by a film badge (or similar dosimeter)
Committed dose equivalent	$H_{T,50}$ (CDE)	Estimated organ dose over the 50 years following intake of radioactivity
Weighting factor	W_T	The weighting factor for organ T is the ratio of the risk of death from stochastic effects (cancer) from irradiation of organ T to the risk of stochastic effects if the same dose was distributed uniformly over the entire body; this value reflects both the relative radiosensitivity of the organ and the risk of fatality from irradiation; for example, the W_T for the thyroid is 0.03 (since thyroid cancer is generally treatable); in contrast, the W_T for bone marrow is 0.12 (due to its high radiosensitivity and the risk of leukemia)[d]
Effective dose equivalent	H_E (EDE)	$H_T \times W_T$ (calculated for all exposed organs and summed)
Committed effective dose equivalent	CEDE	$H_{T,50} \times W_T$ (calculated for all exposed organs and summed)
Total organ dose equivalent	TODE	The total dose to an individual organ from both internal and external exposure = $H_{T,50} + H_d$ (except for lens of eye)
Shallow dose equivalent	SDE	External exposure to the skin or extremity, generally measured by a ring badge
Lens dose equivalent	LDE	External exposure in the vicinity of the eye measured by a film badge
Total effective dose equivalent	TEDE	$H_T + H_{E,50}$

Note:
[a] 10CFR Part 20, Standards for Protection Against Radiation 20.1003 Definitions.
[b] See Chapter 10.
[c] N represents other modifying factors that are not relevant to nuclear medicine exposures; therefore N = 1.
[d] A table of weighting factors is included in 10CFR Part 20, Standards for Protection Against Radiation 20.1003 Definitions.

exposure in an **unrestricted area** (such as the waiting room) must be less than 0.02 mSv/h (2 mrem/h).

Background whole body radiation, at sea level, is approximately 3.6 mSv/year (360 mrem/year), including 2.0 mSv/year (200 mrem/year) from radon.

Methods for Limiting Exposure

Limiting Occupational Exposure

Limiting External Exposure
The three methods of reducing external exposure relate to time, distance, and shielding.

Table 13-2 NRC Radiation Dose Equivalent Limits for Occupational Exposure (Abridged)

Organ/System	Limit	
	rem/year	mSv/year
Total effective dose equivalent (TEDE)[a]	5	50
Total organ dose equivalent (TODE)[a]	50	500
Lens of eye dose equivalent (LDE)[a]	15	150
Shallow dose equivalent to skin or extremity (SDE)[a]	50	500
Dose equivalent to embryo/fetus[b]	0.5 (0.05/month)	5 (0.5/month)
Dose equivalent to minors (<18 years)[c]	10% of adult limit	10% of adult limit

Note:

[a] 10CFR Part 20, Standards for Protection Against Radiation, 20.1201, Occupational dose limits for adults.

[b] 10CFR Part 20, Standards for Protection Against Radiation, 20.1208, Dose equivalent to an embryo/fetus (For workers who have declared their pregnancy).

[c] 10CFR Part 20, Standards for Protection Against Radiation, 20.1207, Occupational dose limits for minors.

Time

Minimize the time spent in the vicinity of a source of radiation. Work efficiently, but do not rush.

Distance

Maintain as large a distance from the source as practical. The radiation intensity from a source (patient or dose) diminishes rapidly as the distance from the source is increased. In Figure 13-1, the bird in the middle is closest to the source (the bird on the left) and receives more radiation than the more distant bird on the right. The radiation dose decreases as the inverse square of the distance (r) from the source ($1/r^2$). Figure 13-2 illustrates areas of equal exposure at distances of (r) and (2r) from a source. At a distance r, the entire dose is spread over a sphere with a surface area of $4\pi r^2$. At twice this distance (2r), the dose is spread over a sphere with four times the area ($16\pi r^2$); the radiation dose at twice the distance (2r) is equal to one fourth of the dose at a distance (r).

When handling radioactive doses, tools such as tongs can effectively reduce exposure to hands and forearms.

Shielding

When time and distance alone are not sufficient, shielding is usually used. Shields take many forms: lead aprons, syringe shields, vial shields, countertop shields (often with leaded glass), fixed and portable (on casters) lead barriers, as well as thinner shields of plastic that may be used for beta-emitting and low-energy gamma-emitting sources. The use of lead or other dense materials as shielding for beta particles is discouraged because the dose will be increased owing to the bremsstrahlung effect (see Chapter 2).

Limiting Internal Exposure

Protection techniques are oriented mainly toward preventing the radioactive material from entering the body. Entrance is most commonly by inhalation, but ingestion, absorption through intact skin, and intake through skin puncture is also possible.

Limiting inhalation is accomplished by good laboratory design, including attention to adequate air replacement and good airflow patterns, use of fume hoods, and by other laboratory practices developed with the consideration

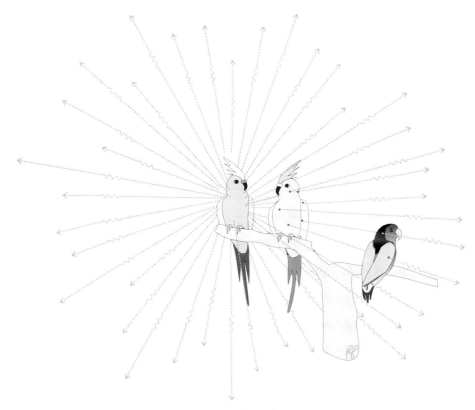

Figure 13-1 Exposure decreases as a function of distance from the source.

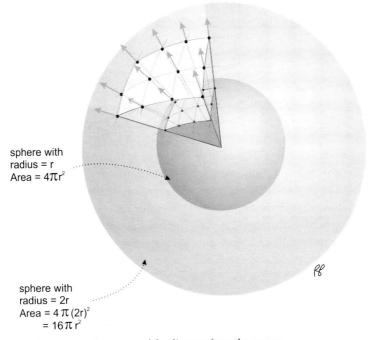

sphere with
radius = r
Area = $4\pi r^2$

sphere with
radius = 2r
Area = $4\pi (2r)^2$
    = $16\pi r^2$

Figure 13-2 Exposure decreases as the square of the distance from the source.

of minimizing inhalation. For example, when working with a volatile radioactive source, the worker should be "up wind" of the source, that is to say, the source should be between the worker and an exhaust such as a hood. Although obvious, this practice is often not followed. The use of respirators to limit inhalation of airborne radioactive material is almost never required in a medical institution and should be considered only as a last resort.

Limiting **ingestion** is accomplished by good laboratory hygiene, such as wearing protective gloves and hospital coats when preparing doses or handling body fluids from a radioactive patient. It is strongly recommended that hands be washed after the removal of gloves. To reduce the risk of inadvertent ingestion of radioactivity, eating, drinking, and smoking in radiation areas is strictly prohibited. Most compounds are not absorbed through intact skin. Because of their chemical composition, however, beta emitters such as ^{32}P and ^{131}I are absorbed readily through the skin. When dealing with these materials extra care should be taken to properly cover the hands, forearms, and other parts that could become contaminated.

Reducing the Risk of Contamination Following a Radiation Spill

Careful handling of radioactive materials can reduce the risk of a **spill**. In the event of a spill it is important to reduce the spread of contamination. Spills are considered minor or major spills depending on the type of pharmaceutical and the estimated amount spilled [1]. Minor spills are controlled by notifying persons in the area that a spill has occurred, cleaning up the spill, and surveying the area. Major spills require more aggressive measures, including clearing the area of non-contaminated individuals, covering, but not cleaning up the spill, marking the boundaries of the spill, and locking the room. In both cases the RSO must be notified of the event [2].

Employer and Employee Responsibilities in Controlling Risk

If an employee is likely to receive more than 10% of any annual limit, the employer is required to perform monitoring. In the case of external exposure, this is usually done with a personal dosimeter such as a film badge or other type of monitor. If the exposure is likely to be internal, for example, from vapors of ^{131}I, monitoring thyroidal uptake is recommended to assess contamination. If the exposure is from ^{32}P, the radiation officer must assess the body burden by measurement of urine counts. A total dose should be estimated based on the effective half-life of the radioactivity.

At least annually, employers are required to notify each monitored employee of the employee's dose. Employers also are required to notify an employee in the event that any limit has been exceeded.

If an employee decides that the risks associated with occupational radiation exposure are too high, the employee may request a reassignment by the employer; however, the employer is not required to provide such a reassignment.

Employees should immediately notify their supervisor if they suspect that a work condition is unsafe or an NRC or state regulation or provision of the license has been violated. The NRC requires licensees to post Form NRC-3, which summarizes employee rights and responsibilities [3].

Limiting Exposure to Patients

Patient doses are calculated with the intention of reducing the exposure to as low a value as possible while performing a clinically useful diagnostic test or treatment. Special consideration must be given to the **pregnant patient**. The benefit of the test for the mother should be weighed against potential risk to the fetus. In general, limited diagnostic testing is possible. The administered dose should be reduced to the lowest feasible value. **Breast-feeding** should be discontinued until there is no significant amount of radiopharmaceutical present in breast milk, which depends on the effective half-life of the radiopharmaceutical. Based on a limited number of available cases, guidelines have been devised for breast-feeding mothers. Breast-feeding can be interrupted for as little as 4 hours for radiopharmaceuticals such as ^{99m}Tc MAA and ^{99m}Tc MDP. A 36-hour cessation is recommended for ^{99m}Tc-pertechnetate, and complete cessation is recommended following administration of compounds such as ^{67}Ga-citrate and ^{131}I-iodide [4].

Limiting Exposure to Family Members and the Public

Education of patients and their family members is important particularly following the administration of beta emitters such as ^{131}I and ^{89}Sr. ^{131}I and ^{89}Sr which excreted through bodily fluids, predominantly via urine. To limit exposure to family members and members of the general public, patients must be carefully instructed on how to reduce contamination. Hygienic precautions, such as flushing the toilet twice after use, hand washing, and separate laundering of clothes and linen are necessary to prevent contact with radioactive urine, sweat, and saliva. In addition, following administration of ^{131}I which emits high energy gammas, patients and family must be educated on the rules of time and distance, as outlined above. Patients and family members should be given written guidelines on the preceding precautions for reference after treatment.

Household contacts of patients receiving radioactivity should not receive more than 5 mSv (500 mrem) total effective dose equivalent from exposure to the patient. For children or pregnant women this limit is 1 mSv (100 mrem) (10 CFR35.75). Until recently, to ensure exposures did not exceed these values, hospital admission was required for doses of sodium iodide

^{131}I that equaled or exceeded 1110 MBq (30 mCi). Patients can now receive higher activities as outpatients if, using dose calculations, it can be shown that the total effective dose equivalent to household contacts will not exceed 5 mSv (1 mSv for children and pregnant). The procedure for calculating doses is outlined in *NUREG-1556* [5].

In general, though, hospitalization is still recommended for incontinent patients, or patients who cannot care for themselves. The hospital rooms used for admission must be designated for radiation therapy and be monitored by the radiation safety staff to reduce staff, visitor, and other patient exposure.

Regulations

The Nuclear Regulatory Commission (NRC) regulations pertaining to the above text and other standard procedures performed in the routine practice of nuclear medicine are summarized in Appendix D.

References

1 U.S. Nuclear Regulatory Commission, Consolidated Guidance about Materials Licenses: Program-Specific Guidance about Medical Use Licenses, *NUREG-1556*, October 2002, 9: Appendix N, Table N.1.
2 U.S. Nuclear Regulatory Commission, Consolidated Guidance about Materials Licenses: Program-Specific Guidance about Medical Use Licenses, *NUREG-1556*, October 2002, 9: Appendix N.
3 Powsner ER and Widman JC, Basic Principles of Radioactivity and its Measurement. In: Burtis C and Ashwood E, eds, *Tietz Textbook of Clinical Chemistry*. 3rd edition, WB Saunders, Philadelphia, 1998.
4 Mountford PJ and Coakley AJ, A Review of the Secretion of Radioactivity in Human Breast Milk; Data, Quantitative Analysis and Recommendations.
5 U.S. Nuclear Regulatory Commission, Consolidated Guidance about Materials Licenses: Program-Specific Guidance about Medical Use

Licenses, *NUREG-1556*, October 2002, 9: Appendix U, Supplement b.

Test Yourself

1 Associate the following annual exposures with the permissible maximum doses:
  (a) TEDE, total effective dose equivalent, for a member of the public
  (b) Maximum permissible exposure to a hospital worker who is not a radiation worker
  (c) Background whole body radiation at sea level
  (d) Maximum permissible TEDE for a radiation worker
    (i) 50 mSv (5.0 rem)
    (ii) 3.6 mSv (0.36 rem)
    (iii) 1.0 mSv (0.1 rem).

2 Select the three most effective ways to reduce exposure when working with radioactivity.
  (a) Reduce the time spent in the vicinity of a radioactive source.
  (b) Use soap when washing your hands.
  (c) Maintain the maximum possible distance from the source.
  (d) Shield the radioactive source.
  (e) Wear a face mask.

3 A nursing mother should be instructed to permanently discontinue breastfeeding under which of the following circumstances?
  (a) After receiving any radiopharmaceutical
  (b) After receiving an injection of 5 mCi of ^{67}Ga citrate
  (c) After receiving an injection of 37 MBq of ^{99m}Tc MAA
  (d) After ingesting 370 MBq of ^{131}I.

4 True or false: The maximum allowable exposure to the household contacts of a patient receiving radioactivity, including children and pregnant women, is 5 mSv.

5 True or false: A tabulated guide to the Nuclear Regulatory Commission regulations is included in the appendices of this book.

Recent events have raised concern about the possibility of a hazardous nuclear event, accidental or purposeful. In any unexpected exposure to ionizing radiation, the person exposed may or may not be aware of the exposure. If the amount of exposure is small the person is unlikely to experience any adverse health effects, and if the person is unaware of the exposure, they will not report to a hospital for treatment. In the unlikely event that an unknown exposure is substantial, the exposed individual may develop generalized symptoms that may be mistaken for another illness. These patients pose a diagnostic dilemma for the health care team and unless the possibility of radiation exposure is considered, the diagnosis will be delayed, if made at all. In the case of a publicized exposure event it is likely that the number of people seeking assistance that were not exposed (but concerned that they were) will be much greater than the number that actually were exposed. Because the number of people that were not exposed can easily overwhelm any health care system designed to treat acutely ill patients, it is important to plan not only for treatment of the injured, but also for triage of the "worried well".

This chapter begins with a brief discussion of the interaction of radiation with tissue. This is followed by a section on facility preparation for decontamination and treatment of victims of a radiation accident. A discussion of clinical techniques for early dose assessment for appropriate triage is followed by a review of the acute radiation syndromes. The chapter concludes with an introduction to the treatment of internal contamination.

Interaction of Radiation with Tissue

Although exposure to neutrons is virtually never encountered in the practice of clinical nuclear medicine, it must be considered a possibility during such events. With this in mind, the discussion of the interaction of radiation with tissue has been expanded to include neutron radiation.

Alpha

Alpha particles have a charge of +2 and a mass of 4 amu. Because they are relatively heavy and charged. They have a short range in matter and a high LET (Chapter 3). They travel only a few centimeters in air and only about 50 μm in tissue. A few layers of dead skin readily block alpha particles and therefore they cannot damage intact skin (Fig. 14-1). On the other hand, the linings of the gastrointestinal and pulmonary system are not covered by dead cells and are susceptible to damage if ingested or inhaled alpha emitters are deposited on them (Fig. 14-2). The high LET of alpha particles means that all of their radiation energy is deposited within these lining cells. Further, rapid cell turnover results in greater total time in mitosis, making the lining cells relatively vulnerable to radiation injury (see Chapter 11).

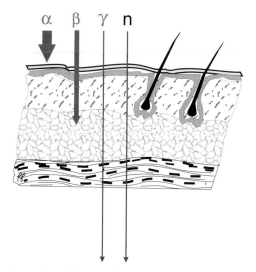

Figure 14-1 Alpha particles are stopped by layers of dead epidermis; high energy beta particles can penetrate short distances into skin; gamma photons and neutrons can penetrate into deep tissue.

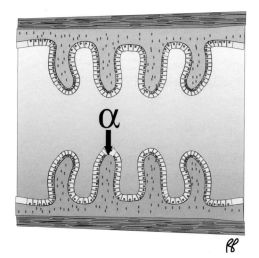

Figure 14-2 Alpha particles can damage the lining epithelium of the gastrointestinal tract or pulmonary system.

Beta

Beta particles have a charge of -1 and a mass of 0.00055 amu. They have a lower LET and penetrate matter further than alpha particles of the same energy; they travel as far as a few meters in air and a few millimeters in tissue. The

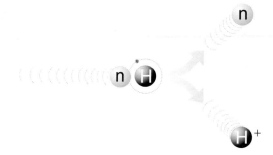

Figure 14-3 Elastic scattering.

higher-energy beta particles are responsible for skin burns. Beta emitters, like all radiation, can also do significant harm if ingested or inhaled in sufficient amounts.

Gamma and X-Ray

Photons have no charge and no mass. As they are uncharged they have relatively few interactions with surrounding matter or air. They can travel several kilometers in air. The greater the energy of the photons the more deeply they penetrate tissue. Exposure to photons may result in both superficial and deep tissue-injury.

Neutrons

Neutrons are different from the preceding types of radiation in a number of ways. They have a finite existence with a half-life of 12 min in air, following which they decay into a proton, electron, and neutrino. They have a mass of 1 amu and have no charge. Since they have no positive or negative charge they lack the great attractive or repellant forces in their interactions with atoms that we have seen with charged particles (alphas and betas). For this reason, as well as for differences in mass, the interactions of neutrons with atoms are limited to the nuclei.

Also, like photons, neutrons can penetrate and pass through tissue; they differ from photons in that they have mass and as a result can interact directly with the hydrogen nuclei in tissue through a process called **elastic scattering** (Fig. 14-3).

Table 14-1 Selected Long-Lived Radionuclides Identified in Previous Radiation Accidents and/or Detected in fallout from Chernobyl

Nuclide	Physical Half-Life	Emissions
^{3}Hydrogen (Tritium)	12 years	$\beta-$
^{60}Cobalt	5.27 years	$\beta-, \gamma$
^{90}Strontium	28 years	$\beta-$
^{131}Iodine	8 days	$\beta-, \gamma$
^{137}Cesium	30 years	$\beta-, \gamma$
^{192}Iridium	74 days	$\beta-$
^{235}Uranium	7×10^8 years	α, γ
^{238}Uranium	4.5×10^9 years	α, γ
^{238}Plutonium	88 years	γ, α
^{239}Plutonium	2.4×10^4 years	γ, α
^{241}Americium	458 years	α

Source: Derived in part from Soloviev V, *et al.* [1].

In this process some of the kinetic energy of the neutron is transferred to the proton and the proton is separated from its atom. The proton becomes a moving charged particle which can damage surrounding tissue. Exposure to neutrons can result in both superficial and deep tissue-injury.

Radionuclides

Some of the longer-lived radionuclides that have been identified in previous industrial radiation accidents and/or in the radioactive fallout from Chernobyl are listed in Table 14-1.

Hospital Response to a Radiation Accident

In the event of a large radiation accident there may be crowds of individuals presenting to the hospitals for evaluation, and it is likely that the majority of these individuals will be uninjured. Screening a large number of otherwise well individuals for contamination and possible radiation exposure can overwhelm limited hospital resources which are needed to treat injured patients. Establishing a **screening and decontamination facility** at an off-site location and redirecting people to that location may be a reasonable solution to this problem. Alternatively an on-site decontamination facility separate from the hospital, such as a large tent, or peripheral building may suffice. A prepared hospital emergency department can then be reserved for medical treatment and decontamination of the injured and critically exposed.

Exposure and Contamination

Radiation accidents can result in partial body or in whole body irradiation. If radionuclides are released during the accident an individual's skin and/or clothing can become **contaminated** (Fig. 14-4A). A contaminated individual is a potential source of radiation exposure for other patients and hospital personnel. An irradiated, non-contaminated person is *not* a source of radiation exposure for others (Fig. 14-4B).

Hospital Facilities

Decontamination Facility
In preparation for a large number of contaminated, but well individuals, many hospitals have designated an area such as a tent or peripheral hospital building with easily controlled access, clothing collection hampers, multiple showers with large waste collection tanks, replacement clean clothing, and several survey meters. These facilities should be staffed by properly clothed (see below) personnel with knowledge of the use of survey meters.

Treatment/Decontamination Room for Seriously Wounded Individuals
The treatment of life threatening wounds or medical conditions takes priority over decontamination and patients with these conditions should

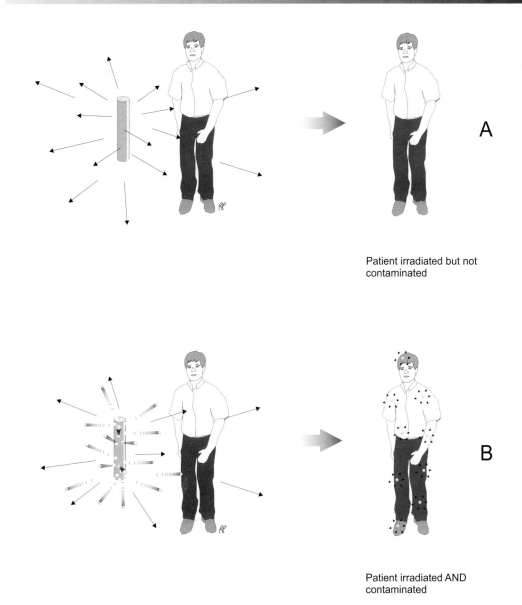

A

Patient irradiated but not
contaminated

B

Patient irradiated AND
contaminated

Figure 14-4 Exposure and contamination.

be immediately transferred to a **treatment/
decontamination room** in the emergency department. Ideally this room should have immediate
access to the outside and should have ambulance
access. A buffer zone should be established for
decontamination and monitoring of personnel
leaving the treatment room prior to entering the
rest of the hospital. If there is adequate notifica-
tion the treatment room may be prepared with
multiple layers of securely taped floor covering
which has a non-skid surface. The purpose of
the floor covering is primarily to aid in clean-
up. Several large plastic lined waste containers
should be present, and any equipment present in
the room which is not to be used should be covered to help prevent contamination. A shielded
area or an area at least 6 feet away from the treatment area should be identified for containment
of radioactive debris such as shrapnel from a
wound. A pair of long handled tongs for handling

metal fragments and survey meters should be available.

After the patient is medically stabilized decontamination can proceed. Although the standard practice of having emergency room or ambulance personnel remove a victim's clothing will often accomplish the majority of the surface decontamination, further cleaning may be required. Wounds can be flushed; however, using a damp washcloth or even the application of minimal water will help to avoid spreading contamination. Abrasion of skin or wounds is contraindicated.

External Decontamination

Undressing removes approximately 90% of radioactive contamination, a damp washcloth or showering removes most of the remaining 10%. Shampooing with gentle shampoos without conditioners prior to skin washing is recommended. Conditioners bind particles to hair. Cleansing should be gentle, abrading the skin may increase absorption of contaminants. and damage to the irradiated skin may delay healing.

Patient Radiation Survey

Following removal of clothing and prior to decontamination the patient should be **surveyed** for baseline readings. When performing surveys it is important to move the probe slowly and to keep a fixed distance from the body. If possible careful records of distribution of contamination and readings at sites of contamination should be recorded. Swabs of contaminated sites should be obtained and labeled with site and time of acquisition.

Following washing, the survey should be repeated using the same distance used for the earlier survey. Decontamination should proceed until the survey readings are twice background or three attempts are unsuccessful and the reduction of radioactivity is less than 10% with each washing. Areas that remain contaminated can be

swathed in plastic to encourage sweating. Sweating will help remove contamination. The plastic should be removed periodically and the area redressed as needed.

Survey Meter

A standard **Geiger-Mueller survey meter** (see Chapter 4) with a **frisking or pancake probe** (Fig. 14-5) is adequate for use in decontamination. The very thin mica window behind the protective wire mesh will allow detection of gamma photons, alpha, and some beta particles.

This method of surveying is limited by the inability of **low energy betas** such as those from tritium to penetrate the mica. Due to their short range in air, **alpha particles** can only be detected with the probe placed very close to area to be surveyed (Fig. 14-6). However, one must not actually touch the surface to avoid contamination of the probe itself.

Covering the probe with a glove to prevent its contamination may also block alpha particles and prevent their detection. Probes are available that are more efficient for the detection of alpha particles and low energy beta particles, but most hospitals do not have these specialized detectors available.

A rough **discrimination of alpha, beta, and gamma emissions** can be determined by comparing the reading directly from the source with that obtained with an intervening piece of paper to block alpha particles and then with an intervening piece of aluminum or plastic to block both alpha and beta particles (Fig. 14-7).

SURVEY METER QUALITY CONTROL

Prior to use a battery check and source check should be performed on the survey meter. The inspection sticker on the side of the meter should be checked to verify that the yearly calibration is up-to-date. For more detailed information on the calibration of survey meters see Chapter 10.

Figure 14-5 Frisking probe for attachment to survey meter.

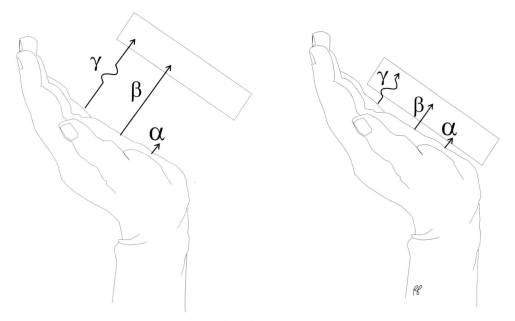

Figure 14-6 Detector proximity is necessary for alpha detection.

Personnel

Preventing Contamination

Personal Protection Equipment

Caps or hoods (particularly for individuals with long hair), eye protection, masks, gowns, double gloves (with the inner glove taped to the gown sleeve), and plastic shoe covers are recommended. Plastic is preferable to paper for shoe covers for durability and resistance to liquids. Gowning and gloving with materials that are readily available in the surgical suite is often perfectly acceptable. Using standard surgical clothing also insures that the clothing needed is always available and in adequate supply.

If there is a credible risk of contamination from radioactive iodine, medical personnel should be treated with oral potassium iodide as soon as possible following the incident. The suggested dose is included in Table 14-3.

Reducing Exposure

The standard practices for minimizing total exposure are reducing the **time** of exposure, increasing the **distance** from the source of the radiation, and when feasible, placing **shielding** between the radiation source and personnel. These remain applicable when treating radiation accident victims. If possible when treating heavily contaminated patients, rotate staff to

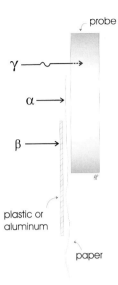

Figure 14-7 Paper and plastic or aluminum can be placed between the radioactive source and the detector to aid in differentiating alpha, beta, and gamma emissions.

reduce time spent with the patient. When not directly caring for a contaminated patient, staff should try to remain at least 6 ft away from the patient. Radioactive debris should be shielded with lead or placed at least six feet from treatment areas. Hospital-based medical personnel treating radiation accident victims, including those who treated victims from the Chernobyl accident, have received only minimal radiation exposures.

Dosimeters

Personnel should be supplied with direct reading dosimeters, such as pocket dosimeters, so that immediate readings are available at the time of exposure. These devices do not provide a means of permanent record keeping and cannot discriminate between types or energies of radiation. All personnel should, therefore, wear film badges or thermoluminescent devices (TLDs) in addition to direct reading dosimeters. For a more detailed description of these devices please refer to Chapter 4 (Figs 4-10 and 4-13 with accompanying text).

Evaluation of Radiation Accident Victim

As soon as possible a complete blood count, blood for **HLA typing**, urine and fecal samples, and nasal swabs should be obtained from a victim of a radiation accident. The **initial lymphocyte counts** can be used with subsequent counts to estimated absorbed dose. HLA typing can be used for future transfusions and stem cell or bone marrow transplants, if needed. The blood, urine, fecal, and nasal samples can be used to estimate quantity and type of internal contamination. A 24 hour (h) post-exposure blood sample should be sent for analysis of **lymphocyte chromosomal abnormalities** for more accurate dosimetry.

The symptoms and medical course of an accident victim are related to the type of radiation, the absorbed radiation dose, the distribution of the absorbed dose (whole body versus localized) and the route of exposure (external, internal). For example, exposure of intact skin, which is protected by several layers of dead skin cells, to a non-absorbable alpha emitter causes relatively little injury whereas ingestion of the same quantity of material can be fatal from damage to the unprotected cells lining the intestines or lungs. Similarly a single exposure resulting in an absorbed dose of 10 Gy to a hand from a gamma emitter will result in severe burns and tissue damage but survival is likely; a whole body dose of 10 Gy from the same emitter will be lethal.

Early Estimation of Whole Body Radiation Exposure

It is important to identify those patients requiring hospital admission, monitoring, and treatment following radiation exposure. Radiation survey measurements from the site of the radiation accident, patient symptoms and time of onset, and blood lymphocyte counts can be used to estimate absorbed doses.

Symptoms and Time of Onset Following Exposure

An absorbed whole body dose greater than approximately 1 Gy may cause **radiation sickness** characterized by specific signs and symptoms. The severity of the signs and symptoms and the rapidity of their onset increase with increasing absorbed dose. The most commonly sought data for early estimation of absorbed dose is the time of onset of **nausea and vomiting** following the radiation accident. Unfortunately, nausea and vomiting are also commonly seen in people involved in a sudden traumatic event that are not exposed to radiation. These symptoms may be caused by abdominal or head trauma or merely by anxiety surrounding the event. It is important that the treating physician be made aware that exposure to radiation occurred so that they can consider the importance of the symptoms.

Table 14-2 summarizes the signs and symptoms of radiation sickness and recommended patient disposition at increasing estimated whole body doses.

Blood Lymphocyte Counts

Neutrophil, platelet, and lymphocyte counts decrease following radiation exposure. Lymphocytes are relatively more radiosensitive and their counts drop more rapidly following exposure. Table 14-3 lists ranges of **estimated doses based on the minimal lymphocyte counts** within the first 48 hrs following exposure. It should be noted that lymphocyte counts can be adversely affected by many factors including stress and underlying illness.

Using data from x-ray, gamma, and eventually neutron/gamma radiation accidents Goans, et al. [5,6] devised a technique of estimating absorbed dose based on lymphocyte counts sampled every 2–3 hrs during the first 6–12 hrs following exposure. These values are plotted using semi-logarithmic paper with time post-exposure (in days) along the non-log x-axis and lymphocyte counts ($\times 10^9$) along the log y-axis. The slope of the initial linear portion of the curve (approximately the first 6 to 8 hrs) can be multiplied by a factor of 8.6 to give an estimated absorbed dose in gray (valid for doses between 50 cGy and 10 Gy). For further details please refer to the articles cited at the end of this chapter.

Chromosomal Aberrations

Tests for quantitation of chromosomal aberrations in lymphocytes are considered to be the most reliable biologic markers for dosimetry measurements. These tests take time to perform and are not widely available. Techniques for more rapid assessment of chromosomal aberrations are in the process of development [7,8].

Early Estimation of Local Radiation Exposure

The early biologic consequences of radiation exposure to a portion of the body or the whole body can be helpful in determining the radiation dose. The time of onset of **initial erythema** of the skin (so-called to distinguish it from the later episodes of more intense reddening of the skin) can be used to estimate absorbed local dose and guide subsequent patient management. Table 14-4 summarizes these findings.

Acute Radiation Sickness

Acute radiation sickness following whole body exposure is often divided into four stages. The **prodromal stage** (or syndrome) includes symptoms of anorexia, nausea, vomiting, and easy fatigability. The greater the dose the shorter the duration of the prodromal stage, but the more severe the symptoms. Immediate diarrhea, fever, headache, and hypertension are seen only in doses above the lethal threshold. During the **subsequent period** the patient may have few or no symptoms in the presence of ongoing organ damage. **Manifest illness** is marked by symptoms related to the involved organ (see syndromes listed below). The **recovery phase** may extend from weeks to years if the radiation dose is not acutely lethal. The overall duration of

Table 14-2 Signs, Symptoms, and Recommended Disposition of Exposed Patients

Estimated Whole Body Dose (Gy)	Onset of Vomiting	Percent of Cases	Diarrhea Severity and Onset	Percent of Cases
<1	none	—	None	—
1–2	>2 h	10–50%	None	—
2–4	1–2 h	70–90%	None	—
4–6	<1 h	100%	Mild 3–8 h	<10%
6–8	<30 min	100%	Heavy 1–3 h	>10%
>8	<10 min	100%	Heavy <1 h	100%

Estimated Whole Body Dose (Gy)	Headache Severity and Time of Onset	Percent of Cases	Fever	Percent of Cases	Level of Consciousness
<1	None	—	None	—	Normal
1–2	Slight	—	None	—	Normal
2–4	Mild	—	Mild 1–3 h	10–80%	Normal
4–6	Moderate 4–24 h	50%	Moderate to High 1–2 h	80–100%	Normal
6–8	Severe 3–4 h	80%	High <1 h	100%	May be reduced
>8	Severe 1–2 h	80–90%	High <1 h	100%	Unconscious—may be for only seconds or minutes [greater than 50 Gy incidence is 100%]

Estimated Whole Body Dose (Gy)	Recommended Disposition
<1	Outpatient with 5 week follow-up of blood labs, skin exams
1–2	Outpatient with anti-emetic therapy, and frequent laboratory and physical exam follow-up or general admission for symptomatic treatment and observation
2–4	Admission with supportive care and hematology consult
4–6	Admission to tertiary care facility with intensive supportive care including fluid management, hematology care, infectious disease consults
6–8	Same
>8	Likely lethal, although with intense medical support may survive up to 12 Gy exposure

Source: Adapted with permission from the International Atomic Energy Agency, Safety Report Series, No. 2, Diagnosis and Treatment of Radiation Injuries [2] with additional information from Cosset, JM, Girinsky, T, Helfre, S, and Gourmelon, P, Medical Management During the Prodromal and Latent Periods [9].

these stages is dependent on the dose the patient received.

Acute Radiation Syndromes

The course of illness following exposure to a single whole body dose of greater than 2 Gy is directly related to the absorbed dose. The Acute Radiation Syndromes may be divided into three categories:

1 Hematopoesis—Approximately 2–10 Gy
2 Gastrointestinal—Approximately 10–30 Gy
3 CNS; Cardiovascular—more than 30 Gy.

Hematopoetic Syndrome

The prodrome consists of nausea, vomiting, diarrhea, headache, fever, and possible decreased level of consciousness presenting within 6 h of exposure and lasting days. During the latent period of 2–3 weeks, the patient will feel relatively well, but the number of circulating red blood cells, white blood cells, and platelets will be steadily decreasing. Following the latent phase, the manifest illness is characterized by infection and hemorrhage.

Gastrointestinal Syndrome

The prodrome consists of the same symptoms of acute radiation sickness seen in the hematopoetic syndrome with nearly universal decrease in level of consciousness beginning within minutes of exposure and lasting days. Latency is approximately one week long, during which time the patient feels relatively well. However, during the latency period the intestinal tract lining is sloughing which leads to the manifest gastrointestinal illness stage when the patient is likely to expire from fluid loss, electrolyte imbalance, and sepsis.

Central Nervous System (CNS) and Cardiovascular Syndrome

The patient will suffer nearly immediate nausea, vomiting, hypotension, ataxia, convulsions, and loss of consciousness. There is no latency period and death is likely within days.

Treatment of Acute Radiation Sickness

Although patients have survived following acute whole body doses of up to 12 Gy, survival

Table 14-3 Estimated Absorbed Dose Based on Minimal Lymphocyte Counts within 48 hours Following Exposure

Estimated Absorbed Dose (Gy)	Lymphocyte Counts (per mm^3)
0–0.4	>1500
0.5–1.9	1000–1499
2.0–3.9	500–999
4.0–7.9	100–499
>7.9	<100

Source: Adapted from Mettler, FA and Voelz, GL, *New England Journal of Medicine*, 2002 [4].

Table 14-4 Estimated Local Radiation Dose and Recommended Patient Management Based on Time of Onset of Early Erythema

Estimated Skin Dose (Gy)	Time of Onset of Initial Erythema and/or Abnormal Skin Sensation	Recommended Patient Management
<6	None[a]	Outpatient with 5 week follow-up of blood labs and exams
6–10	1–2 days[a]	Outpatient with 5 week follow-up of blood labs and exams
8–15	12–24 h	Admission to general ward for observation
15–30	8–15 h	Admission to surgical burn ward with hematology consult
>30	3-6 h, Edema of mucosa can also be seen	Admission to tertiary care facility with intensive surgical (burn) and hematology support

Note:
[a] Mettler, Fred A, *Medical Effects of Ionizing Radiation*, p. 215, Chapter 6.
Source: Adapted with permission from the International Atomic Energy Agency, Safety Report Series, No. 2, Diagnosis and Treatment of Radiation Injuries [2].

following 8 Gy in a single dose is very unlikely without medical treatment. Approximately 50% of those patients exposed to an acute whole body dose of 3.5 Gy will succumb within 2 months without medical treatment; with intensive medical treatment 50% survival may be achieved after exposure to as much as 5.5 Gy [2]. For doses above 2 Gy, hospitalization with isolation and close observation is recommended. The medical management of acute radiation sickness requires a multidisciplinary approach. Specialists in burn care, intensive care, infectious disease, hematology, and radiation safety often need to be closely involved in patient care.

Treatment of Internal Contamination

Patients can become contaminated internally with radiation by ingestion, inhalation, or absorption of specific nuclides through intact skin or open wounds.

Internal contamination with radiation is treated by blocking gastrointestinal absorption, blocking organ specific uptake of the radionuclide, and/or promoting excretion of the contaminants. The treatment is directed at the element and is independent of the specific isotope of that element. For example the treatment of ingestion of all isotopes of strontium is the same. The process of reducing the amount of internal contamination is sometimes referred to as **decorporation**. See Table 14-5 for a listing of internal contaminants and recommended treatments.

Local Radiation Injury to the Skin

The clinical signs of local radiation injury to the skin may be similar to those of thermal burns,

Table 14-5 Recommended Treatment of Internal Contamination with Selected Radionuclides

Name of Radionuclide	Recommended Treatment (Adults Unless Otherwise Indicated)[a]	Treatment Rationale	Remarks
Tritium	Increase oral fluid intake up to 3–4 liters/day, use diuretics[b]	Dilute tritium and increase excretion	Increased fluids and diuretics should be used with caution in patients with renal and cardiac failure
Strontium	Ammonium chloride: 1–2 g orally four times daily up to 6 days[c]	Create a metabolic acidosis	If barium sulfate is used, it should be given immediately
	Sodium alginate: 10 g powder diluted in water, po[c] or	Binds and blocks intestinal absorption	Calcium gluconate should not be given to individuals with a slow heart rate, or
	Barium sulfate: 200–300 g in aqueous suspension po[c]	Binds and blocks intestinal absorption	those taking quinidine or digitalis preparations
	IV calcium gluconate: 2.5 g in 500 ml D$_5$W, IV over 4 hours[c]	Competitive binding in bone	
Iodine	Potassium iodide as soon as possible after incident and daily for duration of exposure Daily dose (po): 130 mg adult 65 mg child (3–18 years) 32 mg child (1 month–3 years) 16 mg infant (<1 month)[d]	Block thyroid uptake	Repeat dosing in pregnant women and infants should be avoided if possible to prevent neonatal hypothyroidism Adolescents approaching adult size ($\geq$ 70 Kg) should receive adult dose of 130 mg

Continued

Table 14-5 Continued

Name of Radionuclide	Recommended Treatment (Adults Unless Otherwise Indicated)[a]	Treatment Rationale	Remarks
Cesium	Insoluble Prussian blue (ferric hexacyanoferrate): Adults: 3 g orally three times daily for 30 days Children (2–12 years): 1 g orally three times daily for 30 days[e]	Enhances excretion	—–
Uranium	Sodium bicarbonate orally or IV: Oral dose: 2 sodium bicarbonate tablets (650 mg each) every 4 hours until urine pH is 8–9 IV dose: 2 ampules sodium bicarbonate (44.3 mEq each, 7.5%) in 1000 ml normal saline at 125 cc/hr[f]	Alkaline urine promotes excretion	—
Plutonium Americium	First 24 hours: Calcium DTPA After first 24 hours: Zinc DTPA Dose for both: 1 g given either by 1 slow IV push over 3–4 minutes, 2 diluted in 100–250 ml D_5W, or NS 3 inhaled in a nebulizer (1 : 1 dilution with water or saline) All doses must be given in less than 2 hours[g]	Chelated radionuclide is more readily excreted	—
Cobalt	Early treatment: gastric lavage and purgatives Severe cases: Penicillamine 250 mg four times daily, can be increased to a total of 4–5 g/day in divided doses[c]	Reduce gastrointestinal absorption Chelate nuclide for excretion	Decorporation of cobalt is inherently difficult, treatment may have limited success. Caution advised when using penicillamine in patients with penicillin allergies

Note:

[a] Prior to initiating recommended treatment pregnancy status and underlying medical illnesses must be considered and for all medications the manufacturer's recommendations as outlined in the package inserts should be followed.

[b] Goans, RE, Update on the Treatment of Internal Contamination, pp. 201–216. In: Ricks, RC, Berger, ME, and O'Hara, FM, eds, *The Medical Basis for Radiation-Accident Preparedness. The Clinical Care of Victims.* Parthenon, New York/London, 2001.

[c] National Council on Radiation Protection and Measurements, NCRP Report No. 65, Management of Persons Accidentally Contaminated with Radionuclides, Washington, D.C., 15 April 1980.

[d] U.S. Department of Health and Human Services, Food and Drug Administration, Center for Drug Evaluation and Research (CDER), Guidance: Potassium Iodide as a Thyroid Blocking Agent in Radiation Emergencies, December 2001, Procedural (http://www.fda.gov/cder/guidance/4825fnl.htm).

[e] Package insert, Radiogardase Prussian Blue Insoluble Capsules, Heyltex Corp, 2004.

[f] Radiation Emergency Assistance Center Training Site (REAC/TS), Oak Ridge Institute for Science and Education (http://www.orau.gov/reacts/internal.htm).

[g] Radiation Emergency Assistance Center Training Site (REAC/TS), Oak Ridge Institute for Science and Education Package, (http://www.orau.gov/reacts/calcium.htm and http://www.orau.gov/reacts/zinc.htm).

but time of onset is dose dependent. The time of onset of the initial erythema, which may be accompanied by pruritis, can be within 24 h following a single dose of more than 10 Gy and is usually non-existent following a dose of less than 6 Gy (see Table 14-4). This initial erythema persists for approximately one week. The manifest illness stage, i.e., intense reddening and pruritis, presents at 2–3 weeks (sooner at higher doses) following exposure. This phase persists for 20–30 days [9].

Two characteristic skin findings following radiation exposure at higher doses are dry desquamation and wet desquamation. **Dry desquamation** is a reddened, dry, flaking, itchy skin due to partial damage of the basal layer. It occurs above a threshold of 8–12 Gy, typically 25–30 days following exposure. **Moist desquamation** is characterized by blistering, redness, pain, and a weeping discharge resulting from the complete damage of the basal layer of the skin. Moist desquamation occurs beginning on day 20–28 at a threshold of 15–20 Gy [2].

References

1 Soloviev, V, Ilyin, L, Baranov, *et al.* Chapter 9. Radiation Accidents in the Former U.S.S.R. In: Gusev, IA, Guskova, AK, Mettler, FA, eds, *Medical Management of Radiation Accidents*, 2nd Edition, CRC Press, pp. 157–165, 2001.

2 International Atomic Energy Agency, Diagnosis and Treatment of Radiation Injuries, Jointly sponsored by the International Atomic Energy Agency and The World Health Organization, Safety Report Series, No. 2, Vienna, 1998 (http://www-pub.iaea.org/MTCD/publications/PDF/P040_scr.pdf).

3 Cosset, JM, Girinsky, T, Helfre, S, and Gourmelon, P, Medical Management During the Prodromal and Latent Periods, pp. 45–51. In: Ricks, RC, Berger, ME, O'Hara, FM, eds, The Medical Basis for Radiation-Accident Preparedness, The Clinical Care of Victims, Proceedings of the Fourth International REAC/TS Conference on The Medical Basis

for Radiation-Accident Preparedness, March 2001, The Parthenon Publishing Group.

4 Mettler, FA and Voelz, GL , Major Radiation Exposure—What to Expect and How to Respond, *New England Journal of Medicine*, 2002, 346(2): 1554–1561.

5 Goans, RE, Holloway, EC, Berger, ME, and Ricks, RC, Early Dose Assessment in Criticality Accidents, *Health Physics*, October 2001, 81(4): 446–449.

6 Goans, RE, Holloway, EC, Berger, ME, and Ricks, RC, Early Dose Assessment Following Severe Radiation Accidents, *Health Physics*, April 1997, 72(4): 513–518.

7 Amundson, SA, Bittner, M, Meltzer, P, Trent J, and Fornace, AJ, Biological Indicators for the Identification of Ionizing Radiation Exposure in Humans. *Expert Rev Mol Diagn*, 2001, 1(2): 89–97.

8 Blakely, WF, Pataje GS, and Miller, AC, Update on Current and New Developments in Biological Dose-Assessment Techniques, pp. 23–32. In: Ricks, RC, Berger, ME, O'Hara, FM, eds, The Medical Basis for Radiation-Accident Preparedness, The Clinical Care of Victims, Proceedings of the Fourth International REAC/TS Conference on The Medical Basis for Radiation-Accident Preparedness, March 2001, The Parthenon Publishing Group.

9 Mettler, Fred A, Medical Effects of Ionizing Radiation, Chapter 6, *Direct Effects of Radiation*, W.B. Saunders and Company, Philadelphia, 1985.

10 Armed Forces Radiobiology Research Institute, *Medical Management of Radiological Casualties Handbook*, April 2003 (http://www.afrri.usuhs.mil).

Test Yourself

1 True or false: Removing clothing and washing with soap and water generally removes 90% of external radioactive contamination.

2 When using a Geiger counter with a pancake probe as a survey meter for unknown contaminants which of the following practices are recommended?

(a) A battery check should be performed prior to use.

(b) Keep the probe at least one meter away from the patient for object being surveyed.

(c) Ascertain that the meter calibration is up to date.

(d) Place a glove over the pancake probe to prevent contamination of the probe.

3 Which of the following clinical findings have been used for an initial radiation dose estimate for an accident victim?

(a) The time of onset of nausea and vomiting

(b) Focal edema

(c) Lymphocyte counts within the first 48 hours

(d) Time of onset of skin erythema

(e) Palpitations

(f) Presence of fever.

4 Associate the following statements with the appropriate acute radiation syndrome:

(a) Predominates at absorbed doses greater than 30 Gy

(b) A 2–3 week latency period during which the patient's blood counts are decreasing

(c) No discernable latency period and death usually occurs within days of exposure

(d) Predominates at absorbed doses of 2–10 Gy

(e) A one week latency followed by fluid loss, electrolyte imbalance, and sepsis

Syndrome:

(i) Hematopoetic syndrome

(ii) Gastrointestinal syndrome

(iii) CNS and cardiovascular syndrome

(iv) Pulmonary-hepatic syndrome.

Recommended reading

Basic Physics and Instrumentation

Cherry, SR, Sorenson, JA, and Phelps, ME, *Physics in Nuclear Medicine*, 3rd edition, Saunders, Philadelphia, 2003.

Saha, GB, *Physics and Radiobiology of Nuclear Medicine*, Springer, New York, 2001.

Single Photon Emission Computed Tomography (SPECT)

Wernick, MN and Aarsvold, Jn, *Emission Tomography, The Fundamentals of PET and SPECT*, Elsevier, Inc., San Diego, CA, 2004.

English, R, *The Principles of SPECT*, 3rd Edition, Society of Nuclear Medicine, Reston, Virginia, 1995.

Positron Emission Tomography (PET) and PET-CT

Wernick, MN and Aarsvold, Jn, *Emission Tomography, The Fundamentals of PET and SPECT*, Elsevier, Inc., San Diego, CA, 2004.

Zaidi, H and Hasegawa, B, Determination of the Attenuation Map in Emission Tomography, *Journal of Nuclear Medicine Technology*, 2003, 44: 291–315.

Siebert, JA, X-Ray Imaging Physics for Nuclear Medicine Technologists. Part 1: Basic Principles of X-Ray Production. *Journal of Nuclear Medicine Technology*, 2004, 32: 139–47.

Siebert, JA and Boone, JM, X-Ray Imaging Physics for Nuclear Medicine Technologists. Part 2: X-Ray Interactions and Image Formation, *Journal of Nuclear Medicine Technology*, 2005, 33: 3–18.

Radiobiology

Casarett, AP, *Radiation Biology*, Prentice-Hall, Englewood Cliffs, NJ, 1968.

Hall, EJ, *Radiobiology for the Radiologist*, 5th Edition, Lippincott Williams and Wilkins, Philadelphia, 2000.

Radiation Dosimetry

Loevinger, R, Budinger TE, and Watson, EE, *MIRD Primer for Absorbed Dose Calculations*, Revised edition, Society of Nuclear Medicine, New York, 1991.

Stabin, MG, Stubbs, JB, and Toohey, RE, *Radiation Dose Estimates for Radiopharmaceuticals*, US Nuclear Regulatory Commission, NUREG/CR-6345, Washington, DC, 1996.

Radiation Protection

Shapiro, J, *Radiation Protection. A Guide for Scientists, Regulators, and Physicians*, 4th edition, Harvard University Press, Cambridge, MA, 2002.

56 Federal Register No. 98, Office of Federal Register, National Archives and Records Administration, Washington, DC, 1991.

International Commission on Radiological Protection, *Radiation Protection*, ICRP publication 26, Pergamon, New York, 1977.

Radiation Accidents

International Atomic Energy Agency, Diagnosis and Treatment of Radiation Injuries, Jointly sponsored by the International Atomic Energy Agency and The World Health Organization, Safety Report Series, No. 2, Vienna, 1998 (http://www-pub.iaea.org/MTCD/publications/PDF/P040_scr.pdf).

Armed Forces Radiobiology Research Institute, *Medical Management of Radiological Casualties Handbook*, April 2003 (http://www.afrri.usuhs.mil).

Mettler, FA and Voelz, GL, Major Radiation Exposure—What to expect and how to respond, *New England Journal of Medicine*, 346(2): 1554–1561, May 16, 2002.

Gusev, IA, Guskova, AK, and Mettler, FA, *Medical Management of Radiation Accidents*, 2nd edition, CRC Press, New York, London, 2001.

Ricks, RC, Berger, ME, and O'Hara, FM, The Medical Basis for Radiation-Accident Preparedness, The Clinical Care of Victims, Proceedings of the Fourth International REAC/TS Conference on The Medical Basis for Radiation-Accident Preparedness, The Parthenon Publishing Group, March 2001.

Appendix A. Common nuclides

Nuclide	Production Method	Half-Life	Decay Mode	$E_{\beta avg}$ (keV)	γ Photon Energies (keV)	Abundance	Decay Product	Half-Life of Decay Product
^{99m}Tc	^{99}Mo–^{99m}Tc	6 hours	IC, γ	—	**140**[a]	0.88	^{99}Tc	2.1×10^5 year/s
					18	0.07		
^{67}Ga	^{68}Zn(d,n)^{67}Ga	78 h	EC, γ	—	**93**	0.38	^{67}Zn	Stable
					185	0.20		
					300	0.17		
					393	0.05		
^{111}In	^{109}Ag(α,2n)^{111}In	67 h	EC, γ	—	23	0.70	^{111}Cd	Stable
					172	0.90		
					245	0.94		
^{131}I	^{235}U(n,f)^{131}I	8 days	β^-, γ	**284.1**	80	0.06	^{131}Xe	Stable
					284	0.06		
					364	0.83		
					637	0.07		
^{123}I	^{121}Sb(α,2n)^{123}I	13 h	EC, γ	—	**159**	0.83	^{123}Te	1.2×10^{13} year
					27	0.72		
					31	0.15		
^{133}Xe	^{235}U(n,f)^{133}Xe	5 days	β^-, γ	110.1	**81**	0.37	^{133}Cs	Stable
					31	0.40		
^{201}Tl	^{203}Tl(d,2n)^{201}Pb	72 h	EC, γ	—	x-ray[b]: **70–80**	0.93	^{200}Hg	Stable
^{99}Mo	^{235}U(n,f)^{99}Mo	67 h	β^-, γ	390.1	140	0.82	^{99m}Tc	6 h
	^{98}Mo(n,λ)^{99}Mo				740	0.14	^{99}Tc	2.1×10^5 year
					41	0.14		
^{89}Sr	^{88}Sr(d,p)^{89}Sr	51 days	β^-, γ	**583.1**	909	100	^{89}Y	Stable
^{32}P	^{32}S(n,p)^{32}P	14 days	β^-,	**735.1**			^{32}S	Stable
^{57}Co	^{56}Fe(d,n)^{57}Co	270 days	EC, γ		**122**	0.86	^{57}Fe	Stable
					136	0.11		
^{81m}Kr	^{81}Rb–^{81m}Kr	13 secs	IT, γ		**190**	0.67	^{81}Kr	2.1×10^6 year
^{18}F	^{18}O(p,n)^{18}F	109 min	β^+	**242.1**			^{18}O	Stable
^{15}O	^{14}N(d,n)^{15}O	124 s	β^+	**735.1**			^{15}N	Stable
^{13}N	^{10}B(α,n)^{13}N	10 min	β^+	**491.1**			^{13}C	Stable
^{11}C	^{11}B(p,n)^{11}C	20 min	β^+	**385.1**			^{11}B	Stable
^{82}Rb	^{82}Sr–^{82}Rb	76 s	β^+	**1409**.1			^{82}Kr	Stable

Source: Browne, E, and Firestone, RB, Table of Radioactive Isotopes. Wiley, New York, 1986.

Note:
[a] Bold print refers to medically important decay emissions for imaging, counting, or therapy.
[b] x-rays emitted following electron capture.

Appendix B. Major dosimetry for common pharmaceuticals

Radiopharmaceutical	Sample Adult Dose Used for Calculation of Dosimetry[a]	Principal Target Organ and Absorbed Dose[b]	Total Body Absorbed Dose or Effective Dose Equivalent[c]
[99m]Tc pertechnetate	370 MBq (10 mCi)	Stomach wall 25 mGy (2.5 rad)	1.4 mGy (0.14 rad)
[99m]Tc DMSA[d] (dimercaptosuccinic acid)	222 MBq (6 mCi)	Renal cortex 51.0 mGy (5.1 rad)	0.9 mGy (0.09 rad)
^{99m}Tc-DTPA[e] (pentetate)	370 MBq (10 mCi)*	Bladder wall* 2-h void: 12 mGy (1.2 rad) 4.8-h void: 27 mGy (2.7 rad)	EDE = 2.9 mSv (290 mrem)
[99m]Tc-Hepatolite[f] (Disofenin)	185 MBq (5 mCi)	Upper large intestinal wall 17.4 mGy (1.7 rad)	0.7 mGy (0.07 rad)
[99m]Tc HMPAO[g] (ceretec)	740 MBq (20 mCi)	Lacrimal glands 51 mGy (5.1 rad)	2.7 mGy (0.27 rad)
^{99m}Tc-MAA[h] (macroaggregated albumin)	148 MBq (4 mCi)	Lungs 8.8 mGy (0.88 rad)	0.6 mGy (0.06 rad)
[99m]Tc MAG3[i] (mertiatide, mercaptoacetyl glycine3)	370 MBq (10 mCi)	Urinary bladder wall 48 mGy (4.8 rad)	0.67 mGy (0.067 rad)
^{99m}Tc-MDP[j] (medronate)	740 MBq (20 mCi)	Bladder wall 2-h void: 26 mGy (2.6 rad) 4.8-h void: 62 mGy (6.2 rad)	1.3 mGy (0.13 rad)
[99m]Tc-Myoview[k]	740 MBq (20 mCi)	Gallbladder wall 24.5 mGy (2.5 rad)	EDE = 6.37 mSv (637 mrem)
[99m]Tc-UltraTag RBC[l] (labeled RBC)	740 MBq (20 mCi)	Spleen 22.0 mGy (2.2 rad)	3.0 mGy (0.3 rad)

Continued

Radiopharmaceutical	Sample Adult Dose Used for Calculation of Dosimetry[a]	Principal Target Organ and Absorbed Dose[b]	Total Body Absorbed Dose or Effective Dose Equivalent[c]
^{99m}Tc-SC[m] (sulfur colloid)	148 MBq (4 mCi)	Liver 14 mGy (1.4 rad)	0.8 mGy (0.08 rad)
^{99m}Tc-Sestamibi[n]	740 MBq (20 mCi)	Upper large intestine 36 mGy (3.6 rad)	3.3 mGy (0.33 rad)
^{99m}Tc-WBC (ceretec)[o]	925 MBq (25 mCi)	Spleen 139 mGy (13.9 rad)	EDE = 15.7 mSv (1570 mrem)
^{57}Co B$_{12}$ [p] (^{57}cobalt cyanocobalamin)	0.018 MBq (0.51 mCi)	Liver 0.65 mGy (0.065 rad)	0.05 mGy (0.005 rad)
^{18}F-FDG[q]	370 MBq (10 mCi)	Bladder wall 61 mGy (6.1 rad) 1-h void: 22 mGy (2.2 rad) 2-h void: 44 mGy (4.4 rad)	EDE = 10.0 mSv (1000 mrem)
^{67}Ga[r] (^{67}gallium citrate)	185 MBq (5 mCi)	Lower large intestine 45.0 MBq (4.5 rad)	13.0 mGy (1.3rad)
^{111}InCl-DTPA[s]	18 MBq (500 μCi)	Spinal cord surface 50.0 mGy (5.0 rad)	0.4 mGy (0.041 rad)
^{111}In-WBC[t]	18.5 MBq (500 μCi)	Spleen 130 mGy (13 rad)	3.1 mGY (0.31 rad)
Sodium Iodide ^{123}I[u]	7.4 MBq (200 μCi)	Thyroid (25% uptake) 240 mGy (24 rad)	0.13 mGy (0.013 rad)
Sodium iodide ^{131}I[v]	0.185 MBq (5 μCi)	Thyroid (25% uptake) 65 mGy (6.5 rad)	0.036 mGy (0.0036 rad)
Sodium iodide ^{131}I[v]	370 MBq (10 mCi)	Thyroid (25% uptake) 130,000 mGy or 130 Gy (13,000 rad)	71 mGy (7.1 rad)
^{32}P[w] (sodium phosphate ^{32}P solution)[w]	185 MBq (5 mCi)	Skeleton 3150 mGy (315 rad)	500 mGy (50 rad)

Continued

Radiopharmaceutical	Sample Adult Dose Used for Calculation of Dosimetry[a]	Principal Target Organ and Absorbed Dose[b]	Total Body Absorbed Dose or Effective Dose Equivalent[c]
^{201}Tl[x] (thallous chloride)	74 MBq (2 mCi)	Testes 61.0 mGy (6.1 rad) Thyroid 46.0 mGy (4.6 rad)	EDE = 27.0 mSv (2700 mrem)
^{133}Xe[y] (xenon gas)	1110 MBq (30 mCi)	Lungs 3.3 mGy (0.33 rad)	0.42 mGy (0.042 rad)

Note:

[a] Actual doses will vary according to manufacturer's recommendations, clinical indications, local practice patterns, and patient's age and condition.

[b] Dose for a healthy organ.

[c] Unless specified as EDE, values given are for total body absorbed dose.

[d] Package insert for MPI DMSA, Medi-Physics, 1993.

[e] Package insert for AN-DTPA, CIS-US, 2003.*dose and dosimetry for IV injection for renal imaging.

[f] Package insert for Technetium Tc99m Disofenin, CIS-US,1999.

[g] Package insert for Ceretec, Amersham, 1990.

[h] Package insert for TechneScan MAA, Mallinckrodt, 2000.

[i] Package insert for TechneScan MAG3, Mallinckrodt, 2000.

[j] Package insert for Draximage-MDP, Drazimage, 1998.

[k] Pacage Insert.for Technetium Tc-99m Tetrofosmin, Amersham Health, 2003.

[l] Package Insert for UltraTag RBC , Mallinckrodt, 2000.

[m] Package insert for CIS-sulfur colloid, CIS-US, 1999.

[n] Package insert for Cardiolite, E.I. duPont de Nemours, 1990.

[o] Package insert for Ceretec, Amersham, 1990.

[p] Package insert Cyanocobalamin, Co 57, Mallinckrodt, 1984.

[q] Pachage Insert for Flourodeoxyglucose F18 injection, Eastern Isotopes,2001.

[r] Package insert for Gallium Citrate Ga 67, Mallinckrodt, 2000.

[s] Package insert for MDI Indium DTPA In 111, Medi-Physics, 1990.

[t] Package insert for Indium In 111 Oxyquinoline Solution, Amersham Health, 2002.

[u] Package insert for Sodium Iodide I 123, Mallinckrodt, 2000.

[v] Package insert for Sodium Iodide I 131, Mallinckrodt, 2000.

[w] Package insert for Sodium Phosphate P 32, Mallinckrodt, 2000.

[x] Package insert for Thallous Chloride Tl 201, Mallinkrodt, 2004.

[y] Package insert for Xenon Xe 133 Gas, Mallinckrodt, 2000.

Appendix C. Sample calculations of the S value

Table C-1 lists terms (variables) that are used to calculate the S value.

EXAMPLE 1

The following is an example of the calculation of the S value when the target organ is the thyroid, and the source organ is also the thyroid, after ingestion of 1 cGy (10.0 mCi) of ^{131}I for treatment of hyperthyroidism. To simplify this calculation, we have assumed that the thyroid concentrates 100% of the dose instantaneously, and we will only use photons and beta particles that contribute the most to the S value. In essence, as shown in Table C-2, an S value is calculated for each particle and photon and these are summed into a total S value.

Table C-1 Terms Used in the Calculation of the S Value

Term	SI Units	Traditional Units	Description
E = energy of emissions	eV	MeV	energy of particle(s) and/or photon(s)
n = number of emissions	1/Bq-s	—	abundance of photons or particles produced during each atom disintegration
K = constant for correction of units in equation	—	—	For the traditional units in this example K = 2.13; for SI units, K = 1
Δ = equilibrium dose constant	kg-Gy/Bq-s	g-rad/μCi-hr	Energy (E) times abundance (n) for each particle or photon times K
ϕ = absorbed fraction	—	—	Fraction of photon or particle energy absorbed by target
m = mass	kg	g	Estimated mass of target organ
Φ = specific absorbed fraction	1/kg	1/g	Absorbed fraction (o) divided by mass of organ (m)
S(target ← source) = S value	Gy/Bq-s	rad/μCi-h	Mean absorbed dose to target organ for each unit of cumulated activity ($\tilde{A}$) in source organ S = ΔΦ

Table C-2 Calculation of S Value for Thyroid as Source and Target

Emission	Mean Number of Emissions/ Disintegration (n)	Mean Energy of Emissions $(\bar{E})$ (MeV)	Equilibrium Dose Constant (Δ) (g-rad/μCi-h)	Absorbed Dose Fraction (ϕ) for Thyroid as Source and Target Organ	Specific Absorbed Dose (Φ) for a 20-g Thyroid g^{-1}	S-Factor $(\Delta * \Phi)$ (rad/μCi-h)
β_1^-	0.0080	0.2893	0.0048	1.0	0.05	2.4×10^{-4}
β_2^-	0.0664	0.0964	0.0136	1.0	0.05	6.8×10^{-4}
β_3^-	0.8980	0.1916	0.3666	1.0	0.05	1.83×10^{-2}
γ_1	0.0578	0.2843	0.0350	0.0310	1.1×10^{-4}	3.85×10^{-6}
γ_2	0.8201	0.3644	0.6366	0.0313	1.1×10^{-4}	7.00×10^{-5}
γ_3	0.0653	0.6367	0.0886	0.0313	1.1×10^{-4}	9.74×10^{-6}
γ_4	0.0173	0.7228	0.0267	0.0305	1.1×10^{-4}	2.90×10^{-6}
					Total S =	$\mathbf{1.93 \times 10^{-2}}$

Source: Data from Weber, DA, Eckerman, KF, Dillman, LT, and Ryman, JC. *MIRD Radionuclide Data and Decay Schemes*. Society of Nuclear Medicine, New York 1989; and Snyder, WS, Ford, MR, Warner, GG, and Fisher, HL. MIRD supplement #3. Estimates of absorbed fractions for monoenergetic photon sources uniformly distributed in various organs of a heterogeneous phantom. Pamphlet 5. *Journal of Nuclear Medicine*, 1969, 10: 13 (Appendix A).

Table C-3 Calculation of S Value for Thyroid as Source and Kidney as Target

Emission	Mean Number of Emissions/ Disintegration (n)	Mean Energy of Emissions $(\bar{E})$ (MeV)	Equilibrium Dose Constant (Δ) (g-rad/μCi-h)	Absorbed Dose Fraction for Thyroid as Source Organ and Kidneys as Target Organs Φ	Specific Absorbed Dose (Φ) for 284-g Kidneys	Individual S-Factors $(\Delta * \Phi)$ (rad/μCi-h) for Each Emission
$\beta-_1$	0.0080	0.2893	0.0048	0.0	0.0	0.0
$\beta-_2$	0.0664	0.0964	0.0136	0.0	0.0	0.0
$\beta-_3$	0.8980	0.1916	0.3666	0.0	0.0	0.0
γ_1	0.0578	0.2843	0.0350	4.23×10^{-5}	1.49×10^{-7}	5.22×10^{-9}
γ_2	0.8201	0.3644	0.6366	5.07×10^{-5}	1.78×10^{-7}	1.13×10^{-7}
γ_3	0.0653	0.6367	0.0886	7.44×10^{-5}	2.62×10^{-7}	2.32×10^{-8}
γ_4	0.0173	0.7228	0.0267	8.03×10^{-5}	2.82×10^{-7}	7.53×10^{-9}
					Total S value[a] =	$\mathbf{1.48 \times 10^{-7}}$

Note:
[a] The calculated total S value of 1.48×10^{27} somewhat overestimates the published value of 1.4×10^{27}.

The total calculated S value of 1.93×10^{-2} is a fair approximation of the published (MIRD pamphlet No. 11) value of 2.2×10^{-2}. Our calculated value is based on a subset of the photons and particles and a gland size of 20 g.

Example 2

The dose to the kidneys from the thyroid in the same case as above is calculated in a similar manner in Table C-3, except, beta particle emissions will not contribute to the absorbed dose in the kidney. All particulate radiation (such as alpha and beta) are called **nonpenetrating** (see Chapter 3); they travel a very short distance and all their energy is absorbed in the source organ. Photons (gamma, x-ray) are called **penetrating** radiation; they generally travel long distances and deposit energy outside the source organ. The ϕ for nonpenetrating radiation (ϕ_{np}) = 1.0 if the target and source organ are the same (as in example above), and $\phi_{np} = 0$ if the target organ is different than the source organ.

Appendix D. Guide to nuclear regulatory commission (NRC) publications

1 Title 10, "Energy", Code of Federal Regulations (10CFR)[1]

Title 10 of the Code of Federal Regulations contains the regulations governing the use of nuclear materials by all individuals and organizations with an NRC license. Three parts of this document are relevant to the practice of nuclear medicine: *Part 19: Notices, Instructions, and Reports to Workers: Inspection and Investigations, Part 20. Standards for protection against radiation, and Part 35. Medical use of byproduct material.*

2 NUREG-1556, Consolidated Guidance About Materials Licenses, Vol. 9, Program-Specific Guidance About Medical Use Licenses[2]

This document is a detailed guide for filling out an NRC license application for the medical use of radionuclides. The appendices I-W contain model procedures for several of the regulations outlined in 10CFR20 and 35.

Sections of these documents relating to routine practice are organized for reference in the following table:

---

[1] Office of the Federal Register, National Archives and Records, Administration, July 1, 2005, Government Printing Office.

[2] Final Report, October, 2002, Prepared by R.W. Broseus, P.A. Lanzisera, A.R. Jones, R.G. Gattone, and R.D. Reid, Division of Industrial and Medical Nuclear Safety, Office of Nuclear Material Safety and Safeguards, U.S. Nuclear Regulatory Commission, Washington, D.C. 20555-0001.

Table D-1 Selected sections of 10CFR Parts 19,20, and 35 and NUREG-1556, Vol. 9

Topic	10CFR Part. Section	NUREG-1556, Vol. 9 Model Procedures
Required posting of NRC Form 3 "Notice to Employees" and 10CFR parts 19 and 20	19.11	
Required instructions to workers concerning risks associated with exposure to radiation	19.12	Appendix J
Requirement for reporting exposure data to individual workers	19.13	
Definitions of terms for Part 20	20.1003	
Develop radiation protection program to comply with regulations and ALARA	20.1101	Appendix N includes model spill clean-up procedures and other Emergency Procedures Appendix T includes rules for wearing gloves, monitoring hands, labeling syringes, etc.
Occupational dose limits for adults	20.1201	
Occupational dose limits for minors	20.1207	
Dose equivalent to embryo/fetus	20.1208	
Public dose limits	20.1301	
Minimum exposure threshold for monitoring individual workers	20.1502	
Requirement for room ventilation or other controls to reduce inhalation of airborne radiation	20.1701 or 20.1702	
Security and surveillance of radioactive materials	20.1801 and 20.1802	
Requirements for posting radiation signs	20.1901, 20.1902	
Requirements for receiving and opening packages	20.1906*	Appendix P
Rules for waste disposal	20.2001	
Record keeping for individual monitoring results	20.2106	
Written directives with procedures for administration	35.40 and 35.41	Appendix S
Training requirements for radiation safety officers, medical physicists, authorized users, nuclear pharmacists.	35.50–35.59	
Use and calibration of the dose calibrator	35.60	
Calibration of survey instruments	35.61	Appendix K
Labeling of vials and syringes	35.69	
End of day surveys	35.70	Appendix R
Criteria for release of individuals following radioactive doses	35.75	Appendix U includes dose calculations for release of patients receiving therapeutic radionuclides
Training requirements for use of radioactive materials for imaging studies for which no written directive is required	35.200	
Training requirements for use of radioactive materials for which a written directive is required	35.300	
Records of written directives and procedures for administration	35.2040 and 35.2041	
Dose calibrator calibration records	35.2060	
Radiation survey instrument calibration records	35.2061	
Records of patient doses	35.2063	
End of day survey records	35.2070	

* 49CFR172.403 defines the transportation indices for radioactive shipping labels. (http://www.gpoaccess.gov/cfr/retrieve.html).

Answers

Chapter 1

1 (1) c, (2) d, (3) e, (4) b, (5) a.
2 b and c are true, a is false; Technetium does not have a stable form; ^{99}Tc has a $T_{1/2}$ of 2.1×10^5 year.
3 e.
4 d.
5 6.3 mCi.

Chapter 2

1 a and c are true, b is false; alpha particles have a shorter range than beta particles.
2 True.
3 False: Compton scattering is the dominant interaction.
4 (a) ii, (b) i, (c) iii.
5 photon interactions: b, c, d, f; charged particle interactions: a, c, e, g.

Chapter 3

1 (e) none of the above, (a) the parent half life is always longer than the daughter half life, (b) if the half life of the parent is between 10 and 100 times greater than the half life of the daughter the activity curve is downward sloping and the equilibrium is termed "transient", and (c) the daughter nuclide is less tightly bound, thereby it can be removed or eluted for use.
2 True.
3 a, b, c, e, f, k are terms for reactors; d, f are terms for cyclotrons; and g, h, i, j are terms for generators.
4 (a) i, (b) i, (c) ii, (d) iii.

Chapter 4

1 False: High-energy photons and x-rays will often pass through the detector without interacting with a gas molecule and low-energy betas may not pass through the detection window.
2 a, c. All of the above are gas detectors, but only the pen dosimeter and the dose calibrator can be classified as ionization chambers. The Geiger counter which functions at a higher voltage range than the ionization chambers is better classified as a gas discharge device.
3 True.
4 True.

Chapter 5

1 (a) iv, (b) iii, (c) ii, (d) i.
2 (a) iv, (b) i, (c) iii, (d) v, (e) ii, (f) vi.
3 True.
4 (a) iii, (b) ii, (c) i.

Chapter 6

1 True. For example, high sensitivity design calls for large holes with thin septa while high resolution calls for small holes and thick septa.
2 True.
3 True.
4 f (4096). In a square matrix, $64 \times 64 = 4096$.
5 f (focusing assembly). This is an undefined term.

Chapter 7

1 (a) ii, (b) i, (c) iv, (d) iii.
2 a and c. Filtering can be applied before, during, and after backprojection so (b) is incorrect.
3 True.
4 (a) low pass filter, (b) high pass filter, (c) high pass filter, (d) low pass filter.
5 e. The Nyquist frequency is 0.5 cycles/pixel, the cutoff frequency refers to the maximum frequency the filter will pass, and the Butterworth filters had the additional parameter called the order of the filter.
6 e.
7 True.

Chapter 8

1 (a) True coincidence event, (b) Random event, (c) Singles event.
2 b and d. High density and short decay are preferable; a and c are incorrect.
3 True.
4 a, c, and d. b is incorrect because septa have nothing to do with 3-D scanning.
5 d.
6 True.

Chapter 9

1 False, 98% of the kinetic energy of the electrons is lost as heat in the target.
2 b and d are correct.
3 d.

4 True.
5 Bone: 1000, Fat: -10, Muscle: 30, Air: -1000, Water: 0.

Chapter 10

1 (a) ii, (b) i, (c) i, (d) iii, (e) i.
2 False: a high count flood, approximately 100 million counts is necessary for SPECT uniformity correction.
3 (a) i, ii, iv, (b) i, ii, (c) i, (d) iii.
4 (a) i, (b) ii, (c) v, (d) vi.
5 False: Deviations greater than one half pixel are abnormal and should be checked with a second collection. A persistent abnormality will require repair prior to further SPECT studies.

Chapter 11

1 b and c.
2 False.
3 True.
4 a, d, b, c.
5 (a) iii, (b) i, (c) ii.

Chapter 12

1 False: The effective half-life is equal to, or shorter than either the physical or biological half-lives.
2 13 days.
3 False: the biological half-life of a radiopharmaceutical is not affected by the physical half-life of the nuclide.
4 True.

Chapter 13

1 (a) iii, (b) iii, (c) ii, (d) i.
2 Select a, c, d.
3 b, d.
4 False: it is 5 mSv for everyone except for children and pregnant women for whom the limit is 1 mSv.
5 True, see Appendix D.

Chapter 14

1 True.

2 a and c are correct. The probe should be used close to, but not touching the skin or object to allow detection of alpha and beta particles. A glove will block detection of alpha particles, so it should not be used to cover the probe.

3 a, c, d, f.

4 (a) iii, (b) i, (c) iii, (d) i, (e) ii.

Index